T0290972

Data Driven Decision Making using Analytics

Computational Intelligence Techniques
Series Editor: Vishal Jain

The objective of this series is to provide researchers a platform to present state of the art innovations, research, and design and implement methodological and algorithmic solutions to data processing problems, designing and analyzing evolving trends in health informatics and computer-aided diagnosis. This series provides support and aid to researchers involved in designing decision support systems that will permit societal acceptance of ambient intelligence. The overall goal of this series is to present the latest snapshot of ongoing research as well as to shed further light on future directions in this space. The series presents novel technical studies as well as position and vision papers comprising hypothetical/speculative scenarios. The book series seeks to compile all aspects of computational intelligence techniques from fundamental principles to current advanced concepts. For this series, we invite researchers, academicians and professionals to contribute, expressing their ideas and research in the application of intelligent techniques to the field of engineering in handbook, reference, or monograph volumes.

Computational Intelligence Techniques and Their Applications to Software Engineering Problems
Ankita Bansal, Abha Jain, Sarika Jain, Vishal Jain, Ankur Choudhary

Smart Computational Intelligence in Biomedical and Health Informatics
Amit Kumar Manocha, Mandeep Singh, Shruti Jain, Vishal Jain

Data Driven Decision Making using Analytics
Parul Gandhi, Surbhi Bhatia, and Kapal Dev

Smart Computing and Self-Adaptive Systems
Simar Preet Singh, Arun Solanki, Anju Sharma, Zdzislaw Polkowski and Rajesh Kumar

For more information about this series, please visit: https://www.routledge.com/ Computational-Intelligence-Techniques/book-series/CIT

Data Driven Decision Making using Analytics

Edited by
Parul Gandhi, Surbhi Bhatia, and Kapal Dev

CRC Press
Taylor & Francis Group
Boca Raton London New York

CRC Press is an imprint of the
Taylor & Francis Group, an **informa** business

CRC Press
Boca Raton and London

First edition published 2022
by CRC Press
6000 Broken Sound Parkway NW, Suite 300, Boca Raton, FL 33487-2742

and by CRC Press
2 Park Square, Milton Park, Abingdon, Oxon OX14 4RN

© 2022 selection and editorial matter, Parul Gandhi, Surbhi Bhatia and Kapal Dev; individual chapters, the contributors

CRC Press is an imprint of Taylor & Francis Group, LLC

Library of Congress Cataloging-in-Publication Data
A catalog record for this title has been requested

ISBN: 978-1-03-205827-6 (hbk)
ISBN: 978-1-03-205828-3 (pbk)
ISBN: 978-1-00-319940-3 (ebk)

DOI: 10.1201/9781003199403

Typeset in Times
by Newgen Publishing UK

Contents

Preface

Digitalization has increased our capabilities for collecting and generating data from different sources. Therefore, tremendous data have flooded in every aspect of our lives. This growth created an urgent need to develop techniques and tools to handle, analyze, and manage data to map it into useful information. This mapping will help improve the performance which eventually supports decision making.

This book brings new opportunities in the area of Data Analytics for Decision Making for further research targeting different verticals such as healthcare and climate change. Further, it explores the concepts of Database Technology, Machine Learning, Knowledge-Based System, High Performance Computing, Information Retrieval, Finding Patterns Hidden in Large Datasets, and Data Visualization. In addition, this book presents various paradigms including pattern mining, clustering and classification, and data analysis. The aim of this book is to provide technical solutions in the field of data analytics and data mining.

This book lays the required basic foundation and also covers cutting-edge topics. With its algorithmic perspective, examples, and comprehensive coverage, this book will offer solid guidance to researchers, students, and practitioners.

Contributors

Contributor's Name	Affiliation	Email
Pradeep Kumar Bhatia	Professor Department of Computer Science and Engineering, Guru Jambheshwar University of Science & Technology, Hisar	pkbhatia.gju@gmail.com
Prerna Bhatnagar	Assistant Professor, Indirapuram Institute of Higher Studies (IIHS), Ghaziabad, Uttar Pradesh	Preranabhatnagar.iihs@gmail.com
Ankur Singh Bist	Chief AI Data Scientist, Signy Advanced Technologies, India	ankur1990bist@gmail.com
Tejinder Pal Singh Brar	Assistant Professor, Department of Computer Applications, CGC Landran, Punjab	tpsbrar@gmail.com3
Kapal Dev	University of Johannesburg, South Africa	kapal.dev@ieee.org
Dagjit Singh Dhatterwal	Assistant Professor, PDM University, Bahadurgarh, Jhajjar, Haryana, India	jagjits247@gmail.com
Parul Gandhi	Professor, Faculty of Computer Applications, MRIIRS, Faridabad	gandhi2110@gmail.com
Ovais Bashir Gashroo	Scholar, Department of Computer Science Jamia Millia Islamia, New Delhi, India	ovaisgashru@gmail.com
Sharad Goel	Director & Professor, Indirapuram Institute of Higher Studies (IIHS), Ghaziabad, Uttar Pradesh	Sharadgoel225@gmail.com
Sonal Kapoor	Associate Professor, Indirapuram Institute of Higher Studies (IIHS), Ghaziabad, UttaPradesh	Sonalkapoor.iihs@gmail.com
Kuldeep Singh Kaswan	Associate Professor, Galgotias University, Greater Noida, Gautam Buddha Nagar, UP, India	kaswankuldeep@gmail.com
Monica Mehrotra	Professor, Department of Computer Science Jamia Millia Islamia, New Delhi, India	drmehrotra2000@gmail.com
Preety	Assistant Professor, PDM University, Bahadurgarh, Jhajjar, Haryana, India	sunnypreety83@gmail.com

Contributor's Name	Affiliation	Email
K. Ramesh	Professor, Department of Computer Science & Engineering, Hindustan Institute of Technology and Science, Chennai, Tamil Nadu, India	kramesh@hindustanuniv.ac.in
Vaibhav Saini	Indian Institute of Technology, Delhi, India	vasa@signy.io
Saima Saleem	Scholar, Department of Computer Science Jamia Millia Islamia, New Delhi, India	saimak6.sk@gmail.com
Ravi Kumar Sharma	Assistant Professor, Department of Computer Applications, CGC Landran, Punjab	ravirasotra@yahoo.com2
D. Sheema	Department of Computer Applications, Hindustan Institute of Technology and Science, Chennai, Tamil Nadu, India	Sheemawilson20@gmail.com
Ashay Singh	Data Scientist, US Tech Solutions Pvt Ltd, India	ashaysingh007@gmail.com

Editors' Biography

PARUL GANDHI

Dr Ghandhi has a is Doctorate in the subject of Computer Science with the study area in Software Engineering from Guru Jambheshwar University, Hisar. She is also a Gold Medalist in M.Sc. Computer Science, with a strong inclination toward academics and research. She has 15 years of academic, research, and administrative experience. She has published more than 40 research papers in reputable international/national journal and conferences. Her research interests include software quality, soft computing, and software metrics and component-based software development, data mining, and IOT. Presently, Dr Gandhi is working as Professor at Manav Rachna International Institute of Research and Studies (MRIIRS), Faridabad. She is also handling the PhD program of the University. She has been associated as an Editorial Board member of *SN Applied Sciences* and also a reviewer with various respected journals of IEEE and conferences. Dr Gandhi has successfully published many book chapters in Scopus-indexed books and also edited various books with well-known indexing databases like Wiley and Springer. She also handles special issues in journals of Elsevier, Springer as a guest editor. She has been called as a resource person in various FDPs and also chaired sessions in various IEEE conferences. Dr Gandhi is the lifetime member of the Computer Society of India.

SURBHI BHATIA PMP®

Dr Bhatia has a Doctorate in Computer Science and Engineering from Banasthali Vidyapith, India. She is currently an Assistant Professor in the Department of Information Systems, College of Computer Sciences and Information Technology, King Faisal University, Saudi Arabia. She has eight years of teaching and academic experience. She is an Editorial board member with Inderscience Publishers in the *International Journal of Hybrid Intelligence, SN Applied Sciences*, Springer, and also in several IEEE conferences. Dr Bhatia has been granted seven national and international patents. She has published more than 30 papers in reputable journals and conferences in well-known indexing databases including SCI, SCIE, Web of Science, and Scopus. She has delivered talks as keynote speaker in IEEE conferences and faculty development programs. Dr Bhatia has successfully authored two books from Springer and Wiley. Currently, she is editing three books from CRC Press, Elsevier, and Springer. She also handles special issues in journals of Elsevier, Springer as a guest editor. Dr Bhatia has been an active researcher in the field of data mining, machine learning, and information retrieval.

KAPAL DEV

Dr Dev is a Postdoctoral Research Fellow with the CONNECT Centre, School of Computer Science and Statistics, Trinity College Dublin (TCD). His education profile revolves over ICT background, i.e. Electronics (B.E and M.E), Telecommunication Engineering (PhD), and Postdoc (Fusion of 5G and Blockchain). He received his PhD degree from Politecnico di Milano, Italy in July 2019. His research interests include blockchain, 5G beyond networks, and artificial intelligence. Previously, Dr Dev worked as 5G Junior Consultant and Engineer at Altran Italia S.p.A, Milan on 5G use cases. He is PI of two Erasmus + International Credit Mobility projects. He is an evaluator of MSCA Co-Fund schemes, Elsevier Book proposals, and top scientific journals and conferences including IEEE TII, IEEE TITS, IEEE TNSE, IEEE JBHI, FGCS, COMNET, TETT, IEEE VTC, and WF-IoT. Dr Dev is TPC member of IEEE BCA 2020 in conjunction with AICCSA 2020, ICBC 2021, DICG Colocated with Middleware 2020, and FTNCT 2020. He is also serving as guest editor (GE) in COMNET (I.F 3.11), Associate Editor in IET Quantum Communication, GE in COMCOM (I.F: 2.8), GE in CMC-Computers, Materials & Continua (I.F 4.89), and lead chair in one of CCNC 2021 workshops. Dr Dev is also acting as Head of Projects for Oceans Network funded by the European Commission.

1 Securing Big Data Using Big Data Mining

Preety[1], Dagjit Singh Dhatterwal[2], and Kuldeep Singh Kaswan[3]
[1]Assistant Professor, PDM University, Bahadurgarh, Jhajjar, Haryana, India
[2]Assistant Professor, PDM University, Bahadurgarh, Jhajjar, Haryana, India
[3]Associate Professor, Galgotias University, Greater Noida, Gautam Buddha Nagar, UP, India

Email ID: sunnypreety83@gmail.com, jagjits247@gmail.com, kaswankuldeep@gmail.com

CONTENTS

DOI: 10.1201/9781003199403-1

1.1 BIG DATA

The advent of IoT (internet of things) devices, business intelligence systems and AI (artificial intelligence) has led to their widespread implementation and to continuously increase the amount of data in existence. The development of self-driving cars, smart cities, home and factory automation, intelligent avionics systems, weaponry automation, medical process automation, Ericsson Company has estimated that nearly 29 billion connected devices are expected by 2022, of which 18 billion would apply to IoT. The number of IoT units, led by the new use scenarios, is projected to grow by 21% between 2016 and 2022. IDC reports that by 2025, real-time data will be more than a quarter of all data. Over the years, control systems kept evolving at different levels of Big Data information security. These control measures although serving as the underlying strategies for securing big data, have limited capability in combating recent attacks as malicious hackers have found new ways of launching destructive operations on big data infrastructures [1].

Digital data will increase as like zettabytes. This forecast gives insight into the higher rate of vulnerabilities and the large scale data security loopholes that may arise. Big data companies are facing greater challenges on how to highly secure and manage the constantly growing data.

Some of the challenges include the following:

- Interception or corruption of data in transit.
- Data in storage which can be held internee by malicious parties or hackers.
- Output data can also be a point of malicious attack.
- Low or no encryption mechanism over the variety of data sources.
- Incompatibility resulting from the various forms of data implementation from different sources.

1.1.1 BIG DATA V's

The above-outlined challenges greatly impact the Vs of big data building blocks that are illustrated in Figure 1.1 [2].

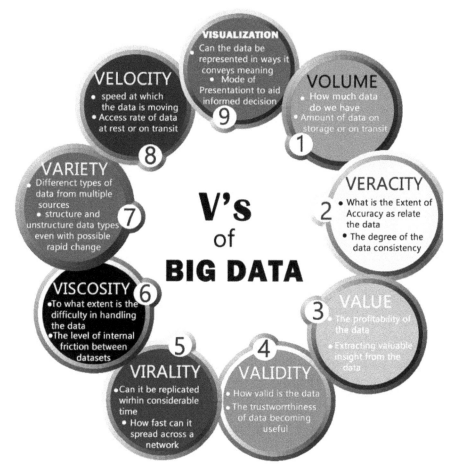

FIGURE 1.1 Nine V's of big data

1.1.1.1 Volume

The cumulative number of data is referred to in the volume. Today, Facebook contributes to 500 terabytes of new data every day. A single flight through the United States can produce 240 terabytes of flight data. In the near future, mobile phones and the data that they generate and ingest will result in thousands of new, continuously changing data streams that include information on the world, location, and other matters.

1.1.1.2 Variety

Data are of various types such as text, sensor data, audio, graphics, and video. Various data forms exist.

Structured data: data that can be saved in the row and column table in the database. These data are linked and can be mapped into pre-designed fields quickly, for example relational database.

Semi-structured data: partially ordered data such as XML and CSV archives.

Unstructured data: data which cannot be pre-defined, for example text, audio, and video files. It accounts for approximately 80% of data. It is fast growing and its use could assist in company's decision making.

1.1.1.3 Velocity

Measuring how easily the data is entering as data streams constantly and receiving usable data in real time from the webcam.

1.1.1.4 Veracity

Consistency or trust of data is veracity.

It investigates whether data obtained from Twitter posts is trustworthy and correct, with hash tags, abbreviations, styles, etc.

- Do you have faith in the data you gathered?
- Is the data enough reliable to gather insight?

1.1.1.5 Validity

It is important to verify the authenticity of the data prior to processing large data sets.

1.1.1.6 Visualization of Big Data

A big data processing task is how the findings are visualized since the data is too broad and user-friendly visualizations are difficult to locate.

1.1.1.7 Value

It refers to the worth of the data being extracted. The bulk of data having no value is not at all useful for the company. Data needs to be converted into something valuable to achieve business gains. Through the estimation of the full costs for the production and processing of big data, businesses can determine whether big data analytics really add some value to their business relative to the ROI that business insights are supposed to produce.

1.1.1.8 Big Data Hiding

Huge volumes of usable data are lost when fresh information is mainly unstructured and dependent on files.

1.1.2 CHALLENGES WITH BIG DATA

- Storing exponentially growing huge data sets.
- Integrating disparate data sources.
- Generating insights in a timely manner.
- Data governance.
- Security issues.

There are so many challenges in handling big data.

1.1.3 ANALYTICS OF BIG DATA

It analyzes the broad and diverse forms of data in order to detect secret trends, associations, and other perspectives.

1.1.3.1 Use Cases Used in Big Data Analytics

1.1.3.1.1 Amazon's "360-Degree View"

In order to develop its recommendation engine, Amazon uses broad data obtained from consumers. It makes recommendations on what you buy, your reviews/feedback, any personal details, your shipping address (to guess your income level based on where you live), and browsing behavior. The company also makes recommendations based on what other customers with similar profile bought. This also helps in retaining their existing customers [3].

1.1.3.1.2 Amazon – Improving User Experience

Amazon is analyzing any visitor clicking on its web pages which will allow the company to understand user's web navigation behavior, their empirical paths to purchase the app, and the paths that led them to leave the site. All this knowledge helps consumers enhance their marketing and advertising experiences.

1.1.4 SOCIAL MEDIA ANALYSIS AND RESPONSE

Companies monitor what people are saying about their products and services in social media, and collect and analyze the posts on Facebook, Twitter, Instagram, etc. This further helps improve their products and enhance customer satisfaction as well as retain existing customers.

1.1.4.1 IoT – Preventive Maintenance and Support

Sensors are used for tracking the system and transmitting the related data over the internet in factories and other installations that use costly instruments. Big data technology programs process to identify whether a crisis is going to occur, often in real time. Prevention of incidents or expensive shutdowns may help its sustain.

1.1.4.2 Healthcare

Big data in healthcare refers to vast volumes of data obtained from a number of sources such as electronic gadgets such as exercise tracking systems, smart clocks, and sensors. Biometric data such as X-rays, CT scan, medical documents, EHR, demographic, family history, asthma, and clinical trial findings also come under big data. It helps physicians develop libraries that are vital to genetic disease prediction. For example, preventive treatment should be carried out for people at risk of a particular illness (e.g. diabetes). According to the data from other human beings, it can include proposals for each patient. Clinical decision support (CDS) software in hospitals investigates on-site diagnostic information and guidance to health providers on diagnosing patients and drafting orders. Wearables continually gather health data from customers and notify physicians. If something goes wrong, the doctor or any other expert will be immediately alerted. Without further interruption, the doctor will call patients and give them all the guidance they need.

1.1.4.3 Insurance Fraud

Insurers analyze their internal data to gain insight into potentially fraudulent claims, call center notices and voice recordings, third party social media reports on bills of individuals, salaries, insolvencies, criminal history, and address change. For example, when a complainant reports about a flood damage to its vehicle, the climatic conditions day can be found out through social media feed. Insurers should apply text analytics to this data that can be find out small inconsistencies found in the case report of the applicant. Fraudsters appear over time to adjust their narrative, making it a valuable tool for the identification of crimes.

1.1.5   Big Data Analytics Tools

1.1.5.1 Hadoop

    a. Hadoop a fully accessible software platform that the Apache Software Foundation developed in 2006. It helps to store data, delivers power to manage big jobs, and carries out virtual simultaneous jobs and tasks. It is intended to incorporate data planning for hustling estimates and shaky lack of activity across figurative hubs [4]. The two fundamental segments include the core which is known as the Hadoop Distributed File System (HDFS) and the processor which is known as the MapReduce motor. HDFS is used to store colossal information which is continually set and duplicated to client application at high transmission capacity, while MapReduce is used for handling gigantic informational indexes in a circulated manner through various machines (Figure 1.2).

The HDFS is Hadoop's stock part. It stores information in fixed blocks by separating the files. Blocks are kept in a cluster of nodes using a master/slave framework in which a cluster contains a single name node and the other nodes are database nodes (slave nodes). NameNode and DataNode are the HDFS Main Components. The HDFS master node retains and handles the blocks in the DataNodes and highly open servers that monitor client's access to files. It also documents metadata for all cluster files such as block location and file size. The files are available in this cluster. Metadata is maintained using the following two files:

- FsImage: contains all the modifications ever made across the Hadoop Cluster NameNode from the beginning (stored on the disk).
- EditLogs: comprises all recent updates – for example the updates in the last 1 hour (stored in RAM).

The NameNode processes the replication factor and collects the Heartbeat (default 3 seconds) and a block report from all DataNodes to ensure live DataNodes. It is also

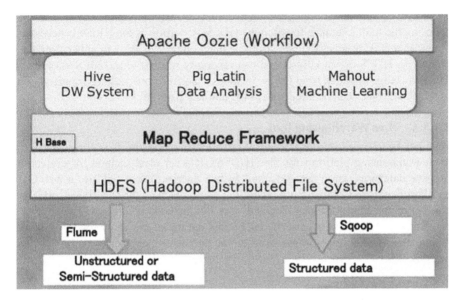

FIGURE 1.2 Hadoop Ecosystem

responsible for choosing new DataNodes in case of DataNode failure. DataNodes behave as slave nodes in the HDFS and store the actual data. DataNodes have a commodity hardware – an unpriced and not good performing device. It serves read and write requests from clients. The secondary NameNode operates along with the main NameNode as a daemon to assist and verify the control. It is used to merge EditLogs with NameNode FsImage and periodically installs EditLogs from the NameNode and adds to FsImage. The next time NameNode begins the new FsImage, it is copied back into the NameNode. It routinely conducts controls (default 1 hour) and hence is also called as CheckpointNode.

1.1.5.2   MapReduce Optimize

This architecture is designed to act as a wide-ranging programming worldview for the processing of a large volume of data on Java-based distributed computers. It includes two capabilities. The first is the Map, which calculates and transforms a number of data into another data set and the Minimize, which results from the Map and treats the tuples in a smaller set of tuples. It is the new step forward, but it is used mostly for estimation. To completely maximize the map reduction process, anything other than calculation must be used. A number of products should be developed so that large knowledge can be handled effectively [5].

1.1.5.3   HBase Hadoop Structure

The Hadoop layout represents a first Bigtable configuration, and the HBase is a mechanism of a hierarchical column-based database used to provide random data access while a large number of organized data is accessed. HBase is primarily responsible for using the fault tolerance found in the HDFS. A portion of the HBase is necessary to ensure that it retrieves the lines from its main register. Users can either store data directly in HDFS or via HBase, and can randomly read/write in HDFS via HBase. The knowledge obtained from HBase includes a lack of properties, key, appreciation, to which attributes can be alluded to in the non-key parts [6].

1.1.5.4   Hive Warehousing Tool

Hive has been based on the Hadoop Distributed File System. It is considered as a great warehousing platform for files (HDFS). It is an ideal method for evaluating massive databases, large data sets, and ad hoc queries. Users will use a web GUI and Java Database Compatibility (JDBC) to communicate with this framework. The concept of the Map Minimize demands that job methods be created. The structure can be viewed as a core part of the HDFS and the top of the knowledge distributor. Applications and constant transactions on the internet are not processed on the Hive network [7].

1.1.5.5   Pig Programming

The Hadoop ecosystem is another really powerful method that provides an additional database for better efficiency. The Pig table is a tuple assemble in which a number of tuples are worth each field. The language of the procedure data flow is called pig Latin and is used primarily for programming. The language provides all the fixed SQL definitions, including entering, organizing, projecting, and collecting samples. Compared with the MapReduce scheme, it also makes for a greater extraction standard, since a Pig Latin inquiry can be modified for a succession of MapReduce firms [8].

1.1.5.6   Mahout Sub-Project Apache

In 2008, Apache's Lucene was created as a sub-project and is an open-source platform used mainly to create usable algorithms for machine learning. The following methods of machine language are used:

1.   Collaborative Sifting
2.   Clustering
3.   Categorization/Classification [9]

1.1.5.7   Non-Structured Query Language

It is a non-SQL or non-relational database; it is a database that provides a framework for the recovery and preservation of non-social databases of data. Different kinds of NoSQL databases, which represent a report for key pairs of values, parts, and map databases, software engineers, demonstrate that their pre-owned applications have reasonable details about the structure. Because of the growth of simple internet use and free storing of information, a massive amount of ordered, semi-organized, and non-structured information is obtained and everything is set away for different kinds of uses. Normally, this information is intended as large information. You're using NoSQL databases for Google, Twitter, Amazon, and other famous websites [10].

1.1.5.8 Bigtable

A Bigtable system was introduced in 2004 and is currently used by many Google users such as Map Minimize. They are also used in Bigtable, Google Readers, Google Maps, Google Book Search, Google Earth, Blogger.com, Google Code facilitations, Orkut, YouTube, and Gmail to provide and alter information. Encouragement Google has flexibility and greater control of execution on particular database. The wide table is expanded using scattered data stock piling the board model, which relies on the stock of the parts in order to boost data recovery.

1.1.6 SECURITY THREATS FOR BIG DATA

- An unauthorized customer can access documents and perform authentication or other attacks.
- An unauthorized customer can snap/spy parcels of information to the customer.
- An unauthorized client will view/compose a record details square.
- An unauthorized consumer can gain advantages to corrupt important information in the related files.

1.1.7 BIG DATA MINING ALGORITHMS

Big data implementation requires free sources and decentralized controls; it is entirely exorbitant, owing to the imaginable delivery costs and protection issues for each distributed datum source to the incorporated mining facility. In a double standard procedure, input, model, and information can be stressed all the more clearly in world mining companies. At the level of knowledge, each location will include data measurements depending on the sources of information for the community and share perspectives between directions to achieve a worldwide distribution. Each site will complete neighborhood mining training activities at models or design levels to find neighborhoods models with regard to restricted details. Quickly developing examples can be synthesized by conglomerating designers through all the local sources by exchanging designs between different sources [11]. The model relationship analysis at the level of information shows the importance among models generated from different information sources in deciding how important information sources are connected and how accurate choices that rely on models from self-governing resources can be formed. As they are limited, there are unreasonably not enough information focusers to decide accurate ends. Based on speculation, information is an exceptional category of information fact, where not every information set depends on any simple/incorrect conveyance at this point. Different factors such as the dissatisfaction of a sensor center or certain typical approaches to intentionally skirt some qualities may deliver the missing qualities. Although most knowledge mining measurements today include solutions to tackle missing qualities, the attribution of data is an integrated area of study that aims to construct upgraded models by crediting missing qualities. Complex and dynamic mining data generate motivated information by the exponential development of structured data and its scale and nature progressions. Complex papers are absolutely illustrated with complex details on WWW servers, Website flips, interpersonal institutions, communications systems, and transport systems. Although complex dependency structures depending on the information make our acceptability criteria complexity, however, broad data complexity is applied from a wide variety of viewpoints including increasingly complex types, complicated simple semitone relationships, and complex information association structures.

Big data contains ordered information, unstructured information, semi-structured information, etc. Social databases, texts, hypertexts, pictures, sound and video data are particularly available. Web news, Facebook commentaries, Picasa photographs, and YouTube video on an honorary scholastic occasion. There is no doubt that these details include sound semantical relationships. Mining a dynamic semantic interaction with knowledge "text image video" would primarily lead to developing application systems such as the search engines or suggestion architectures.

There are relations between individuals with respect to big data. In the internet, there are blogs, and hyperlinks are used to create a mind-boggling structure for web pages. Social links also occur between individuals who build complicated social networks, for instance massive Facebook, Twitter, LinkedIn relationship details and other internet-based life, including call detail records (CDRs), gadgets and sensor information, GPS and geocoded map data, huge image documents transferred through the Manager File Transfer Convention, and web texts. Expanding research activities have begun to solve problems of institutional development, swarms engagement, and documentation and communication in agreements of dynamic relationship systems.

1.1.8 BIG DATA MINING FOR BIG DATA SECURITY

The field of data science involves a comprehensive analysis of the vast volume of data, including the retrieval, using scientific techniques, diverse technology, and algorithms, of useful insights from raw, organized, and unstructured data. It uses methods and strategies to exploit data to uncover occult phenomena and develop something unique and important from the knowledge surrounding us. With data science techniques on big data, we can mitigate the rising risk and vulnerabilities on a huge amount of data in our global tech ecosystem, thus implementing highly resilient intelligent security measures from the data science features of predictive analysis, statistical computations, machine learning, and deep learning facilities (Figure 1.3).

Data science advanced analytic systems avail the following:

- Security Information and Event Management (SIEM)
- Security Metrics
- Vulnerability Assessment
- Risk Modeling
- Government Compliance and Risk (GRC) Automation [12]
- Computer and Network Forensics
- File Integrity Scanning

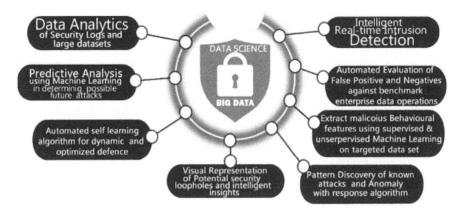

FIGURE 1.3 Data sciences for big data security

1.1.8.1 Securing Big Data

By applying data science analytics and a variety of machine learning tools, we can carry out thorough security analysis of a collection of data to reveal trends and patterns for actionable intel as to what security measures to deploy. An instance could be a system security expert finding that all threats on corporate data takes place at night, or at a certain stage in the day, over a network or offline. He can learn enough to restrict the likelihood to a specific network terminal. In addition, the derived information can be used to forecast future possible attacks. During reconnaissance queries and movements, most likely traces and signals are left behind. Data contains these signals which can be observed by means of data science in order to boost early warnings [13]. All the data were presumably sent to a security data lake named SIEM before development in data science (security information and event management). However, recent data science developments will now lead to associations over many incidents in real time. It is easier to connect dots and identify patterns using algorithms which, due to the lack of security analysts, were previously difficult to find manually. The opportunity to benefit from decision making by security researchers and overtime taking proactive steps and actions as a physical security researcher is part of several benefits of data-based science programs.

1.1.8.2 Real-Time Predictive and Active Intrusion Detection Systems

Hackers and destructive attackers are using a variety of techniques and types to obtain access to massive data stores. These programs track users and devices by data science pattern discovery on the network and flag risky behaviors. The essence of an attack and its magnitude ratio and the future effect degree can be calculated using a data science. Data science increases the application of these methods and simplifies these methods. Such methods can identify possible problems and attack patterns specifically by applying real-time and historic data to a machine learning algorithm. With time, a device like this becomes knowledgeable and precise; potential threats can be forecast and different loopholes found.

1.1.8.3   Securing Valuable Information Using Data Science

Most attacks on big data infrastructure and resources are to perpetuate the loss of extremely valuable data and information, which of course becomes harm to a business. Unauthorized users will avoid checking the data set using authentication mechanisms such as extremely complex signatures or encryption. Impenetrable protocols can be developed through data science. One can, for example, build algorithms to identify the most common target chain of data by studying the background of the cyber-attacks and offline intrusion attempts on organizational data, thus extracting insight on why such data is targeted and the probable outcome. This helps to define appropriate security measures required to be implemented for that focus data [14].

1.1.8.4   Pattern Discovery

Big data security requires data science techniques to play important roles in the present and future generation of defense strategies [15]. Within enterprise data stores and networks, high data sources of enormous volumes are in existence which can facilitate the discovery and prevention of attacks and other malicious offline and network activities. At the very core, the focus is on innovating the probability model-based and statistical approach by utilizing these data sources for identifying subtle intrusion attempts. The post-detection of irregularities is one of the most valuable applications of data science, which has proved its usefulness as soon as the attackers enter the corporate data center; they appear like registered users that use legitimate authorization for accessing systems and data stores. Data science self-learning models enable anomaly detection by helping scientists to understand "normal" behavior inside the enterprise and observe the slightest deviation. The advantage brought by using a data science approach is the ability to pre-learn; from historic data, complex patterns of normal computer and network behavior could be deduced, thereby detecting anomalies which would not stand out otherwise; one such scenario is the unusual network traversal using legitimate credentials. In consideration of data pre-breach context, data science is contributing to enabling big data enterprises identify malicious code and mistaken transactions using supervised engine training technologies surrounded by trees environment to remove behavioral functionality from suspect executables. The businesses that employ this method will no longer use handwriting malware sensing to discourage malicious code writers from identifying them, making it even more complicated. Pre-break using monitored learning and post-break identification of deviations or predictive approaches are fields of data science contributions in terms of big data security [12].

1.1.8.5   Automated Detection and Response Using Data Science

Data science contributes to describing the magnitude of an attack using automated methodologies. Detection and response work side by side; therefore, the more accurate detailing the extent of an attack on any big data stockpile, the better an accurate speedy response can be made. Data science is enhancing automated response which is dependent on the capability of effective detection systems [16]. Training intelligent systems to be more certain by applying a combination of analytics and machine learning on detection before deploying automated response which may be based on false positives is where data science is strengthening the security response suitable for big data protection within an enterprise [17].

1.1.9 Conclusions

As the data volume grows every day, big data will begin to expand and become one of the fascinating prospects in the next few years. We have an incredible increase in volume, speed, and variety of data today. Security and secrecy have also seen exponential development in the vulnerability ecosystem and data protection threats. Now we are in a new era where by using Cloud Computing we can manage all our data with less money and effort, because we are taking the route of outsourcing for managing big data [18]. Not only we can manage with less effort but we can manage big data in a very effective manner using MapReduce technique. Through effective data science analytics, machine learning, and statistical computations systems, the security of big data can be greatly enhanced beyond the already existing control systems. This can give rise to highly intelligent data security implementation with the ability to self protect and automate actions when a suspicious pattern is detected. IT professionals will be able to advance operative, defensive, and active measures to prevent malicious attacks on big data by the vast options data science presents.

REFERENCES

[1] Bhatia, S. A. (2020). Comparative Study of Opinion Summarization Techniques. IEEE Transactions on Computational Social Systems, 8(1), 110–117.

[2] Kshatri, S. S., Singh, D., Narain, B., Bhatia, S., Quasim, M. T., & Sinha, G. R. (2021). An empirical analysis of machine learning algorithms for crime prediction using stacked generalization: An ensemble approach. *IEEE Access*, *9*, 67488–67500.

[3] Vinodhan D, Vinnarasi A. (2016). IOT based smart home. International Journal of Engineering and Innovative Technology (IJEIT), 10, 35–38.

[4] Kamalpreet Singh, Ravinder Kaur (2014). "Hadoop: Addressing Challenges Of Big Data," IEEE International Advanced Computing Conference (IACC).

[5] Cloudera.com (2015). Introduction to Hadoop and MapReduce [online]. Available at: www.cloudera.com/content/cloudera/en/trainin g/courses/udacity/mapreduce.html (accessed on 16/1/2015).

[6] Charu C. Aggarwal, Naveen Ashish, Amit Sheth (2013). *"The Internet of Things: A Survey from the Data-Centric Perspective, Managing and Mining Sensor Data*, Springer Science+Business Media, New York, pp. 384–428.

[7] Obaidat, M. S., & Nicopolitidis, P. (2016). *Smart Cities and Homes: Key Enabling Technologies*. Cambridge, MA: Morgan Kaufmann, pp. 91–108.

[8] Eugene Silberstein, "Automatic Controls and Devices"*Residential Construction Academy HVAC*, Chapter 7, pp. 158–184.

[9] Scuro, Carmelo & Sciammarella, Paolo & Lamonaca, Francesco & Olivito, R. & Carnì, Domenico. (2018). IoT for structural health monitoring. *IEEE Instrumentation and Measurement Magazine*. 21(6). 4–9 and 14.

[10]   Alojail, M., & Bhatia, S. (2020). A novel technique for behavioral analytics using ensemble learning algorithms in E-commerce. IEEE Access, 8, 150072-150080.

[11]   Sheikh, R. A., Bhatia, S., Metre, S. G., & Faqihi, A. Y. A. (2021). Strategic value realization framework from learning analytics: a practical approach. Journal of Applied Research in Higher Education.

[12]   Bisht B., & Gandhi P. (2019) "Review Study on Software Defect Prediction Models premised upon Various Data Mining Approaches", INDIACom-2019, 10th INDIACom 6th International Conference on "Computing For Sustainable Global Development" at Bharti Vidyapeeth's Institute of Computer Applications and Management (BVICAM).

[13]   Gandhi P., & Pruthi J. (2020) Data Visualization Techniques: Traditional Data to Big Data. In: *Data Visualization*. Springer, Singapore. pp. 53–74.

[14]   Sethi, J. K., & Mittal, M. (2020). Monitoring the impact of air quality on the COVID-19 fatalities in Delhi, vol-1 using machine learning techniques. *Disaster Medicine and Public Health Preparedness*, 1–8.

[15]   Chhetri B et al., Estimating the prevalence of stress among Indian students during the COVID-19 pandemic: a cross-sectional study from India, *Journal of Taibah University Medical Sciences*, pp. 35–50 https://doi.org/10.1016/j.jtumed.2020.12.0

[16]   Dagjit Singh Dhatterwal, Preety, Kuldeep Singh Kaswan, *The Knowledge Representation in COVID-19* Springer, ISBN: 978-981-15-7317-0

[17]   Chawla, S., Mittal, M., Chawla M., Goyal, L. M. (2020). Corona Virus - SARS-CoV-2: an insight to another way of natural disaster. *EAI Endorsed Transactions on Pervasive Health and Technology*, 6, 22.

[18]   Gandhi K., & Gandhi P. (2016) "Cloud Computing Security Issues: An Analysis", INDIACom-2016 10th INDIACom 3rd International Conference on "Computing for Sustainable Global Development" at Bharti Vidyapeeth's Institute of Computer Applications and Management (BVICAM), pp. 7670–7673.

2 Analytical Theory
Frequent Pattern Mining

Ovais Bashir Gashroo[1] and Monica Mehrotra[2]
[1]Scholar, Department of Computer Science,
Jamia Millia Islamia, New Delhi, India
[2]Professor, Department of Computer Science,
Jamia Millia Islamia, New Delhi, India

CONTENTS

2.1   INTRODUCTION

Frequent pattern mining (FPM) is a process in which the association between distinct items in a database needs to be found. Frequent patterns are actually itemsets, sub-structures, or sub-sequences which appear within a set of data and have frequency not below a user-specified limit. For example, collection of items, like bread and eggs,

DOI: 10.1201/9781003199403-2

that look often together in a transaction data set can be called as frequent itemset [1]. A sub-sequence like purchasing a laptop followed by a camera, afterwards, an external hard drive, if it occurs in the history of shopping database often, it will be termed as a frequent sequential pattern. Structural forms like subgraphs or subtrees which can be joined with itemsets or sub-sequences are referred to as sub-structures. And the frequent occurrence of sub-structures within a graph database is known as frequent structural pattern.

FPM has been a requisite task in data mining, and researchers have been a lot more dedicated on this concept from the last couple of years. With its profuse use in data mining problems such as classification and clustering, FPM has been broadly researched. The advent of FPM into real-world businesses has led to the promotion of sales which resulted in increase in profits. FPM has been applied in domains like recommender systems, bioinformatics, and decision making. The literature dedicated to this field of research is abundant and has achieved tremendous progress such as the development of efficient and effective algorithms aimed at mining of frequent itemsets. FPM is of immense importance in many important data mining tasks like association and correlation analysis, analyzing patterns in spatiotemporal data, classification, and cluster analysis.

The process of FPM can be specified as follows: If we have a database DTB which contains transactions T1, T2 ... Tn, all the patterns P need to be determined which appear in no less than a fraction s of all the transactions [2]. Fraction s is usually referred to as the "minimum support". This was put forward first by Aggarwal et al. [3] in 1993 for analyzing market basket as a kind of association rule mining. FPM was able to analyze the buying habits of customers by discovering the associations among the items which are being placed by customers in their respective baskets used for shopping. For example, customers who are buying bread, what are their chances of buying eggs? Information like this can help increase the sales because the owners will do marketing as per this information and shelf spaces will be arranged accordingly.

This chapter will provide a comprehensive study in the field of FPM. The chapter will also explore some of the algorithms for FPM, analysis of FPM algorithms, privacy issues, various applications of FPM, and some of the resources that are available for those who want to practice the FPM methods. Finally, the chapter will be concluded with future directions in this area.

2.2 FREQUENT PATTERN MINING ALGORITHMS

Many researchers have come up with a lot of algorithms to enhance the FPM process. This section analyzes several FPM algorithms to give us an understanding. Generally, FPM algorithms can be categorized under three categories [2], namely Join-Based, Pattern Growth, and Tree-based algorithms. Join-Based algorithms use a bottom-up method to recognize frequent items in a data set and keep on enlarging them into itemsets as long as those itemsets become more than a minimum threshold value described by the user in the database. On the other hand, Tree-Based algorithms use a set-enumeration technique for solving the frequent itemset generation problem. It achieves this through the construction of a lexicographic tree enabling the mining of items via a collection of ways as the breadth-first or depth-first order. Lastly, Pattern Growth algorithms apply a divide-and-conquer method to partition and project databases depending on the presently identified frequent patterns and expand them into longer ones in the projected databases [4]. Algorithms such as Apriori, DHP, AprioriTID, and AprioriHybrid are classified under Join-Based Algorithms, while algorithms such as AIS, TreeProjection, EClaT, VIPER, MAFIA, and TM are classified under Tree-Based Algorithms [4]. FP-Growth, TFR, SSR, P-Mine, LP-Growth, and Extract are classified under Pattern Growth algorithms [4]. Some of the algorithms from all the three categories will be discussed here in this chapter.

2.2.1 APRIORI ALGORITHM

Apriori [5] is the first algorithm which has been used for mining of frequent patterns. This algorithm mines itemsets which are frequent so that Boolean association rules can be generated. An iterative stepwise technique of searching is employed to find (k+1)-itemsets from k-itemsets. An example of transactional data is shown in Table 2.1; it contains items purchased in different transactions. Initially, the whole database is searched so that all frequent 1-itemsets are identified after calculating them. Then only those among them which fulfill the minimum support threshold are captured. The entire database has to be scanned for identifying each frequent itemset and it has to be made sure that no frequent k-itemset is possible to be identified. Supposing minimum support count is 2, that time in our example, records having support greater or equal to minimum support are going to be included into the next phase for processing by the algorithm.

The size of candidate itemsets is lessened significantly through the Apriori algorithm and a significant performance gain is also provided in many of the cases. However, there are few limitations which are critical in nature that this algorithm suffers from [6]. One of them being that if there is a rise in the total count of frequent k-itemsets, large amount of candidate itemsets have to be produced. In response, the algorithm has to scan the entire database recurrently and the verification of a large set of candidate items is needed using the pattern matching technique.

TABLE 2.1
Example of Transactional Data

Transaction ID	Item ID's
T100	a, b, e
T101	b, d
T102	b, c
T103	a, b, d
T104	a, c
T105	b, c
T106	a, c
T107	a, b, c, e

The main benefit of the Apriori algorithm is that it uses an iterative step-wise searching technique for discovering (k+1)-itemsets from k-itemsets. The disadvantages of the Apriori algorithm are that it needs to generate a number of candidate sets when the itemsets are greater in number and also the database has to be scanned repeatedly to determine the support count of the itemsets [4].

2.2.2 DHP ALGORITHM

DHP stands for the Direct Hashing and Pruning method [7] and it was put forward after the Apriori algorithm. Mainly two optimizations are proposed in this algorithm to speed up itself. First is pruning the candidate itemsets in each iteration, and second is trimming the transactions so that the support counting procedure becomes more effective [2].

For pruning the itemsets, the DHP algorithm keeps track of the incomplete information regarding the candidate (k+1)-itemsets, meanwhile counting the support explicitly of candidate k-itemsets. When the counting of candidate k-itemsets is being done, all of the (k+1) subsets are discovered and are hashed in a table that preserves the counting of the number of subsets which are hashed into each entry [2]. The retrieval of counts from the hash table for each itemset is done during the stage of counting (k+1)-itemsets. As there are collisions in the hash table, the counts are overestimated. Itemsets which have counts under the user stated support level are actually pruned for further attention.

Trimming of transactions is the second optimization which is proposed in the DHP algorithm. An item can be pruned from a transaction if it doesn't occur in no less than k frequent itemsets in Fk because it can't be used for support calculation so that frequent patterns can be found any longer. This can be understood by an important observation that if an item doesn't occur in no less than k frequent itemsets in Fk, then any other frequent itemset will be containing that item in it. The width of the transaction is reduced, and the efficiency in terms of processing is increased through this step [2].

2.2.3 FP-GROWTH ALGORITHM

It is known as frequent pattern growth algorithm [6] and it mines frequent itemsets without any costly process for the generation of candidates. The algorithm employs a divide and conquer approach for compressing the frequent items into an FP-Tree which holds all the data related to the association of the frequent items. The division of FP-Tree into a number of conditional FP-Trees for every frequent item is done so that each frequent item can be mined individually [4]. The representation of frequent items with FP-Tree is displayed in Figure 2.1.

The problem of identification of long frequent patterns is being solved in the FP-Growth algorithm by repeatedly searching via conditional FP-Trees which are smaller in size. Examples of conditional FP-Trees and the detailed conditional FP-Trees which are present in Figure 2.1 can be accessed in [8]. As per [4], "The Conditional Pattern Base is a 'sub-database' which consists of every prefix path in the FP-Tree that co-occurs with every frequent length-1 item. It is used to construct the

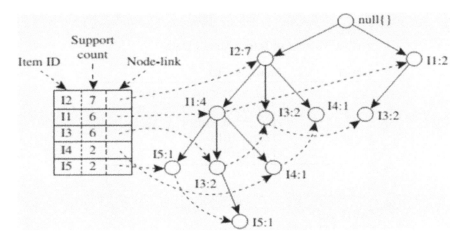

FIGURE 2.1   FP-Tree

Source: Reproduced from [8].

Conditional FP-Tree and generate all the frequent patterns related to every frequent length-1 item". There is a significant reduction for searching the frequent patterns in terms of cost. According to [9], constructing an FP-Tree is a time-consuming process if the set of data available is huge.

The first advantage of this algorithm is that the association related information of every itemset is being preserved and second the volume of data which has to be searched shrinks [3]. However, the disadvantage of this algorithm is that the time required for constructing an FP-Tree will be high if data upon which it is being built is very large [4][9].

2.2.4   ECLaT ALGORITHM

By using the data format, which is vertical in nature, the EClaT (Equivalence Class Transformation) algorithm [10] efficiently mines the frequent itemsets as shown in Table 2.2. Transactions containing particular itemsets are grouped into the same record using this method of data representation. The transformation of data into the vertical format from the horizontal format is done using the EClaT algorithm after looking into the database once. The production of frequent (k+1)-itemsets is accomplished through the intersection of the transactions containing the frequent k-itemsets. This process is being repeated till all the frequent itemsets are intersected with each other and there remains no frequent itemset that can be discovered as depicted in Tables 2.3 and 2.4.

The EClaT algorithm need not scan the database multiple times so that (k+1)-itemsets are identified. For transforming the data from the horizontal to the vertical format, the database needs to be scanned only one time. The support count for each itemset is just the total number of transactions that hold that particular itemset, so the database does not need to be scanned more than once for identifying the support

TABLE 2.2
Vertical Format of Transactional Data

Itemset	Transaction ID
a	{T100, T103, T104, T106, T107}
b	{T100, T101, T102, T103, T105, T107}
c	{T102, T104, T105, T106, T107}
d	{T101, T103}
e	{T100, T107}

TABLE 2.3
Vertical Data Format of 2-Itemsets

Itemset	Transaction ID
{a, b}	{T100, T103, T107}
{a, c}	{T104, T106, T107}
{a, d}	{T103}
{a, e}	{T100, T107}
{b, c}	{T102, T105, T107}
{b, d}	{T101, T103}
{b, e}	{T100, T107}
{c, e}	{T107}

TABLE 2.4
Vertical Data format of 3-Itemsets

Itemset	Transaction ID
{a, b, c}	{T107}
{a, b, e}	{T100, T107}

count. A lot of memory space will be needed and the processing time will be also higher for the intersection of the itemsets if the transactions involved in an itemset are higher [4]. The advantage of the EClaT algorithm is that the database needs to be looked into only once [4]. However, the intersection of long transactions sets takes both more space in memory and longer processing time, which is a disadvantage of this algorithm [4].

2.2.5 TREE PROJECTION ALGORITHM

This algorithm [11] mines frequent itemsets so that a lexicographic tree can be constructed via different techniques of searching like depth-first, breadth-first,

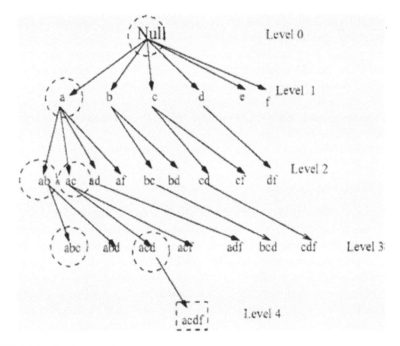

FIGURE 2.2   Lexicographic tree

Source: Reproduced from [11].

or a mixture of both of them. The nodes of the lexicographic tree are used for keeping the support of every frequent itemset in each transaction after it has been calculated by this algorithm. The performance involved in the calculation of the total number of transactions containing a particular itemset is improved [4]. The lexicographic tree representing the frequent items is shown in Figure 2.2 as an example here.

The algorithm within the hierarchical structure of the lexicographic tree will search only those subsets within the transactions which have the chance to hold the frequent itemsets and this is an advantage of the Tree Projection algorithm because the identification of the frequent itemsets is done much faster. A top-down technique is used for traversing the lexicographic tree while searching it. The disadvantage of this algorithm is that it is not efficient in consuming memory space when there are different representations of the lexicographic tree [12].

2.2.6   TM ALGORITHM

Just like the EClaT algorithm, the TM (Transaction Mapping) [13] algorithm also mines the frequent itemsets by making use of the vertical representation of data. For every itemset, this algorithm transforms and maps the transaction IDs in a list of transaction intervals which are present in other positions. After this, using the depth-first search method, intersection is performed among the transaction intervals all the way through the lexicographic tree to facilitate the counting of itemsets. The example of this technique can be visualized in [13].

The compression of transaction IDs in a continuous transaction interval is notably done using the TM technique if there is a high value of minimum support. The advantage of this algorithm is that the time for intersection is saved when the itemsets are being compressed in a list of transaction intervals [4]. For data sets containing short frequent patterns, the TM algorithm achieved performance much better than the FP-Growth and EClaT algorithms [4]. While comparing it with the FP-Growth algorithm, the TM algorithm is slower when we consider the processing speed [4].

2.2.7   P-Mine Algorithm

The P-Mine [14] algorithm is capable of frequent itemset mining on a processor having multiple cores by employing a parallel disk-based approach. The production of dense data set on the disk is done in a less time using a data structure named VLDBMine. The VLDBMine data structure uses a Hybrid-Tree for storing the whole data set and the information that is required for the data retrieval process. The disk access is slow which results in low performance, but here a pre-fetching method is implemented which can load a number of projections of the data set into separate cores of the processor such that the frequent itemsets are mined; this technique enhances the efficiency in terms of disk access. At last, every processor core gives results, and they are finally combined for the construction of the entire frequent itemsets. The P-Mine algorithm architecture is illustrated in [14].

Improvements in performance and scalability are there for frequent itemset mining because of the data set that is represented in the VLDBMine data structure. These improvements are because of the Hybrid-Tree that is used in the VLDBMine data structure. The optimization of scalability and performance when the mining of frequent itemsets is done parallelly with the multiple cores present in a processor is an advantage of the P-Mine algorithm. However, the disadvantage of this algorithm is that the maximum level of optimization can be achieved only when there are multiple cores present in a processor.

2.2.8   Can-Mining Algorithm

The mining of frequent itemsets is done in an incremental way from a Can-Tree (Canonical-Order Tree) using the Can-Mining [15] algorithm. A header table is used by this algorithm that consists of all the database items just like the FP-Growth algorithm. Every item and their corresponding pointers to the first and last nodes which contain the item in the Can-Tree are stored into the header table. A list containing the frequent items is needed so that the algorithm can obtain the frequent patterns from the Can-Tree by performing the mining operation. The advantage of this algorithm is that at times when there is a high threshold value of the minimum support, the Can-Mining algorithm outperforms the FP-Growth algorithm. However, the disadvantage is that the time taken to mine is longer when the minimum support holds a much lower threshold value. The architecture of the Can-Mining algorithm is illustrated in [15].

TABLE 2.5
Timeline of the FPM Algorithms

S. No	Algorithm	Publication Year
1.	Apriori [5]	1994
2.	DHP [7]	1995
3.	FR-Growth [6]	2000
4.	EClaT [10]	2000
5.	Tree Projection [11]	2001
6.	TM [13]	2006
7.	P-Mine [14]	2013
8.	Can-Mining [15]	2015

2.3 ANALYSIS OF THE ALGORITHMS

Some of the recent and important algorithms for mining frequent patterns have been discussed above. Both advantages and disadvantages of these algorithms were discussed. The recent algorithms have to deal with the new kind of data and face problems that the earlier algorithms didn't have to. The overall performance of the algorithms with respect to the execution time and the amount of memory space being used is the key focus in this area. If the algorithm implemented in real-world application doesn't efficiently produce results on time, then it will cause loss to the business where it was used. The main purpose of developing these efficient algorithms is to have less execution time so that results can be produced quickly, which, in turn, will lead to better decision making and eventually increase the sales in markets and help grow businesses. In [9], the author has illustrated the runtime of different algorithms using tables and graphs and also the memory usage of several existing FPM algorithms. Table 2.5 shows the algorithms discussed above along with their year of publication.

2.4 PRIVACY ISSUES

Privacy issue has become a very concerning thing in recent years because the individual's personal data is available widely [16]. The data shared is often reluctant, is shared in a very constrained manner, or a low-quality version of data is shared. These issues pose a challenge in discovering frequent patterns from the data. Below we have discussed the challenges faced in terms of frequent pattern and association rule mining:

- Once methods of privacy preservation like randomization are imposed, discovering association rules from the data becomes a challenge. This is because of the addition of large amount of noise to the data and discovering association rules when noise is present in the data is a difficult task. These class association rule mining methods [17] put forward an efficient method for the meaningful discovery of patterns while maintaining the privacy of the modified data.

- The problem of distributed privacy preservation [18] is that the data which needs to be mined is being kept in a dispersed manner by the market contestants who compete with each other. They want themselves data to find out the global knowledge without revealing their local insights [2].

2.5 APPLICATIONS OF FPM

This section will emphasize mainly on the application part of FPM because it serves as the main motivation for frequent pattern algorithms. These applications span over a variety of other fields incorporating many other domains of data. As we have limited space here, we will be focusing mainly on some of the key areas where it is being used or we can say the application part of FPM.

Some of the key applications of FPM have been discussed in the following.

2.5.1 FOR CUSTOMER ANALYSIS

Supermarket and customer analysis [3] were the initial research put forth by researchers. In this situation, the behavior of customers like what basket of items they purchased at the same time or what sequences of items were purchased together is recorded. What are the common patterns of buying behavior can be answered by using the frequent patterns. Two-tuple frequent pattern like {Milk, Bread} is an example that suggests that the items Milk and Bread are bought together often. With this information the shelves containing these two items can be placed together and also this information can be used to promote items in a much better way. This information helps making the decisions efficiently and producing good results for the business. Because if there is information regarding the previous buying behavior of customers, the decision of keeping things in a store can be made in a much better way. If the previous information regarding a person's buying things is known and after analyzing earlier period information, the marketing process would become easy. For example, if there is already information available regarding a customer who has bought a laptop, it is very much likely that he will buy a printer. So, in that case targeting a customer for a certain item would become easy.

2.5.2 FREQUENT PATTERNS FOR CLASSIFICATION

The problem of data classification is linked to FPM, especially in the rule-based methods context. The condition of the form is known as a classification rule [3].

$$A1 = a1, A2 = a2 => C = c \qquad (2.1)$$

Attributes A1 and A2 will be taking the values a1 and a2, respectively, as implied by the LHS of the rule and value c should be the class value as implied by the RHS of the rule.

Classification rules have a similarity in form as association rules and suitable patterns can be determined from the data with the help of mining techniques of association rules. The aim is to ensure that the patterns are enough discriminative for the purpose of classification and the criteria of support are not having much dominance in the process of rule selection. This [19] is the earliest work that highlights the connection between classification and association rule mining. Classification based on associations (CBA) [20] is one among the most popular methods that classify on the basis of associations. Also, the CMAR [9] method is another technique which is used for classifying based on the FP-Growth method for mining association rules.

2.5.3 Frequent Patterns Aimed at Clustering

Problems in the field of data mining like clustering are related to FPM. In [21], the relationship between clustering and FPM is talked much about, in which the item having a large size is used to enable the clustering process.

The article in [22] and the chapter on high-dimensional data in [23] contain a thorough analysis about the associations among high-dimensional clustering algorithms and FPM problems.

2.5.4 Frequent Patterns for Outlier Analysis

For binary and transaction data, FPM is used often for outlier analysis. The inherent nature of transaction data as high-dimensional needs to identify the relevant outliers present in it. For this subspace, methods are utilized for the identification of outliers. The challenge faced by the subspace methods that they are not computationally practical or statistically feasible so that they can define the sets of items (or subspaces) that are sparse for outlier detection has been addressed within [24].

2.5.5 Frequent Patterns for Indexing

Apart from the clustering methods, the FPM methods are effective in creating representations which the clustering methods [25][26] do in the table of transactions. Clustering methods partition the database containing transactions in groups based on the broad patterns present in them. The use of an FPM approach can be inherent for the market basket data context.

The gIndex [27] method is an indexing structure that uses frequent patterns which are discriminative. There are other methods that have been developed keeping in mind the context like Grafil [28] and PIS [29].

2.5.6 Frequent Patterns for Text Mining

In the field of text mining, frequent patterns have very important applications, for both positional and non-positional co-occurrence [2]. When some words happen to be together in terms of adjacency of occurrence it is referred to as positional co-occurrence. Sequence pattern mining methods can be adapted, or constrained FPM methods can be applied for discovering such patterns. Non-positional co-occurrence is associated with the problem of discovering bigrams, trigrams, or phrases within the data which appears frequently.

For text collections, there are many applications of FPM which are present within [30].

2.5.7   FREQUENT PATTERNS FOR SPATIAL AND SPATIOTEMPORAL APPLICATIONS

There has been a lot of advancements in the field of mobile sensing technology. Social sensing [31] is an important branch that has emerged out of it. A lot of data is accumulated from cell phones uninterruptedly and its large portion is that of data related to GPS, which is nothing but data about the position. The construction of trajectories can be done through this GPS-based location data. The determination of clusters and frequent patterns can be done out of these constructed trajectories.

The clustering of spatiotemporal data has been achieved through the frequent use of FPM methods. The swarm method which has been proposed in [32] is an example of this technique.

2.5.8   APPLICATIONS IN CHEMICAL AND BIOLOGICAL FIELDS

Both biological and chemical data can usually be denoted by graphs. Chemical compounds can be denoted by graphs in which the nodes will be representing the atoms and the bonds in between atoms can be shown as the edges between the nodes of a graph. In a similar fashion, biological data can be shown with graphs or sequences in many ways. In case of biological data, there will be diversity in terms of the structural representations. For the identification of useful and important properties in these chemical compounds or structures, FPM will play a significant role. For classifying chemical compounds an approach is discussed in [33]. Biological data is available as sequence or graph-structured data [2]; useful frequent patterns can be found out from them with the help of many algorithms developed [34]–[38].

2.6   RESOURCES AVAILABLE FOR PRACTITIONER

Because the FPM methods are utilized so often for a lot of applications, we can use software that is available immediately and aimed at providing services to use FPM in these applications.

KDD Nuggets [39] is a website that contains links to different resources on FPM. The implementation of different data mining algorithms which contains algorithms for FPM is present in the Weka repository [40]. Bart Goethals has implemented some of the distinguished FPM algorithms like Apriori, EClaT, and FP-Growth [41]. FIMI is a repository [42] very well known for the efficient implementation of many FPM algorithms. There is one free package from software R called arules which is capable of doing FPM for many types, the details of which can be accessed in [43]. Commercial software like Enterprise Miner which is provided by SAS has the ability to perform both associations and sequential pattern mining [44].

2.7 FUTURE WORKS AND CONCLUSION

Among the main challenges of data mining like classification, clustering, outlier analysis, and FPM, the challenge of FPM is considered to be the leading problem in the domain of data mining. This chapter gives a summary of some of the key areas within FPM. This chapter also reviewed the strong points and weak points of different algorithms. In addition, we also highlighted the issues which arise because of the growing privacy concerns. Applications like analyzing the customer's buying behavior, outlier analysis of data, mining of textual data, and chemical and biological data were discussed along with other applications.

We will be focusing on how to deal with the privacy preservation and effectively mine our data without letting anything happen to the privacy. Also, some more advanced FPM algorithms will be analyzed with their different versions. The focus will be on how the earlier developed algorithms can be modified so that they can work effectively in today's environment. In addition, the variants of FPM shall be dealt with to know how they impact the industry.

REFERENCES

[1] Han J, Cheng H, Xin D, Yan X (2007) Frequent pattern mining: current status and future directions. Data Min Knowl Discov., 15(1), 55–86. https://doi.org/10.1007/s10618-006-0059-1

[2] Aggarwal CC (2014) An introduction to frequent pattern mining. Freq. Pattern Min., 1–17.

[3] Agrawal R, Imieliński T, Swami A (1993) Mining Association Rules Between Sets of Items in Large Databases. ACM SIGMOD Rec. https://doi.org/10.1145/170036.170072

[4] Chee CH, Jaafar J, Aziz IA, et al (2019) Algorithms for frequent itemset mining: a literature review. Artif. Intell. Rev., 52(4), 2603–2621.

[5] Agrawal R, Srikant R (1994) Fast algorithms for mining association rules in large databases. In: Proc. of the 20th International Conference on Very Large Data Bases (VLDB'94).

[6] Han J, Pei J, Yin Y (2000) Mining Frequent Patterns Without Candidate Generation. SIGMOD Rec (ACM Spec Interes Gr Manag Data). https://doi.org/10.1145/335191.335372

[7] Park JS, Chen MS, Yu PS (1995) An Effective Hash-Based Algorithm for Mining Association Rules. ACM SIGMOD Rec. https://doi.org/10.1145/568271.223813

[8] Han J, Kamber M, Pei J (2012) Data Mining: Concepts and Techniques.

[9] Meenakshi A (2015) Survey of frequent pattern mining algorithms in horizontal and vertical data layouts. Int J Adv Comput Sci Technol, 4(4).

[10] Zaki MJ (2000) Scalable algorithms for association mining. IEEE Trans Knowl Data, 12(3), 372–390. https://doi.org/10.1109/69.846291

[11] Agarwal RC, Aggarwal CC, Prasad VVV (2001) A tree projection algorithm for generation of frequent item sets. J Parallel Distrib Comput., 61(3), 350–371. https://doi.org/10.1006/jpdc.2000.1693

[12] Aggarwal CC, Bhuiyan MA, Hasan M Al (2014) Frequent pattern mining algorithms: a survey. In: Frequent Pattern Mining.

[13] Song M, Rajasekaran S (2006) A transaction mapping algorithm for frequent itemsets mining. IEEE Trans Knowl Data Eng., 18(4), 472–481. https://doi.org/10.1109/TKDE.2006.1599386

[14] Baralis E, Cerquitelli T, Chiusano S, Grand A (2013) P-Mine: Parallel itemset mining on large datasets. In: Proceedings of International Conference on Data Engineering.

[15] Hoseini MS, Shahraki MN, Neysiani BS (2016) A new algorithm for mining frequent patterns in can tree. In: Conference Proceedings of 2015 2nd International Conference on Knowledge-Based Engineering and Innovation, KBEI 2015

[16] Aggarwal CC, Yu PS (2008) A general survey of privacy-preserving data mining models and algorithms. In: Privacy-preserving Data Mining (pp. 11–52).

[17] Evfimievski A, Srikant R, Agrawal R, Gehrke J (2004) Privacy preserving mining of association rules. In: Information Systems.

[18] Clifton C, Kantarcioglu M, Vaidya J, et al (2002) Tools for privacy preserving distributed data mining. ACM SIGKDD Explor Newsl. https://doi.org/10.1145/772862.772867

[19] Ali K, Manganaris S, Srikant R (1997) Partial classification using association rules. Knowl Discov Data Min., 97, 115–118.

[20] Liu B, Hsu W, Ma Y, Ma B (1998) Integrating classification and association rule mining. Knowl Discov Data Min., 98, 80–86. https://doi.org/10.1.1.48.8380

[21] Wang K, Xu C, Liu B (1999) Clustering transactions using large items. In: International Conference on Information and Knowledge Management, Proceedings

[22] L. P, E. H, H. L (2004) Subspace clustering for high dimensional data: a review. SIGKDD Explor Newsl ACM Spec Interes Gr Knowl Discov Data Min

[23] Aggarwal CC, Reddy CK (2013) DATA Clustering Algorithms and Applications.

[24] He Z, Deng S, Xu X (2002) Outlier detection integrating semantic knowledge. In: Lecture Notes in Computer Science (including subseries Lecture Notes in Artificial Intelligence and Lecture Notes in Bioinformatics).

[25] Aggarwal CC, Wolf JL, Yu PS (1999) A new method for similarity indexing of market basket data. SIGMOD Rec (ACM Spec Interes Gr Manag Data). https://doi.org/10.1145/304181.304218

[26] Nanopoulos A, Manolopoulos Y (2002) Efficient similarity search for market basket data. VLDB J., 11(2), 138–152. https://doi.org/10.1007/s00778-002-0068-7

[27] Yan X, Yu PS, Han J (2004) Graph indexing: a frequent structure-based approach. In: Proceedings of the ACM SIGMOD International Conference on Management of Data.

[28] Yan X, Yu PS, Han J (2005) Substructure similarity search in graph databases. In: Proceedings of the ACM SIGMOD International Conference on Management of Data.

[29] Yan X, Zhu F, Han J, Yu PS (2006) Searching substructures with superimposed distance. In: Proceedings of the International Conference on Data Engineering.

[30] Aggarwal CC, Zhai CX (Eds.) (2013) Mining text data. Springer Science & Business Media

[31] Aggarwal CC, Abdelzaher T (2013) Social sensing. In: Managing and Mining Sensor Data (pp. 237–297).

[32] Li Z, Ding B, Han J, Kays R (2010) Swarm: mining relaxed temporal moving object clusters. In: Proceedings of the VLDB Endowment, 3(1). https://doi.org/10.14778/1920841.1920934

[33] Deshpande M, Kuramochi M, Wale N, Karypis G (2005) Frequent substructure-based approaches for classifying chemical compounds. IEEE Trans Knowl Data Eng., 17(8), 1036-1050. https://doi.org/10.1109/TKDE.2005.127

[34] Cong G, Tung AKH, Xu X, et al (2004) FARMER: finding interesting rule groups in microarray datasets. In: Proceedings of the ACM SIGMOD International Conference on Management of Data.

[35] Cong G, Tan KL, K.h.tung A, Xu X (2005) Mining top-k covering rule groups for gene expression data. In: Proceedings of the ACM SIGMOD International Conference on Management of Data.

[36] Pan F, Cong G, Tung AKH, et al (2003) Carpenter: Finding closed patterns in long biological datasets. In: Proceedings of the ACM SIGKDD International Conference on Knowledge Discovery and Data Mining.

[37] Wang J, Shapiro B, Shasha D (1999) Pattern Discovery in Biomolecular Data: Tools, Techniques and Applications. Oxford University Press.

[38] Rigoutsos I, Floratos A (1998) Combinatorial pattern discovery in biological sequences: the TEIRESIAS algorithm. Bioinformatics, 14(1), 55–67. https://doi.org/10.1093/bioinformatics/14.1.55

[39] KDnuggets www.kdnuggets.com/software/associations.html

[40] www.cs.waikato.ac.nz/ml/weka/

[41] http://adrem.ua.ac.be/ ~goethals/software/

[42] http://fimi.ua.ac.be/

[43] https://cran.r-project.org/web/packages/arules/index.html

[44] www2.sas.com/proceedings/forum2007/132-2007.pdf

3 A Journey from Big Data to Data Mining in Quality Improvement

Sharad Goel¹ and Prerna Bhatnagar²
¹Director and Professor, Indirapuram Institute of Higher
Studies (IIHS), Ghaziabad, Uttar Pradesh
²Assistant Professor, Indirapuram Institute of Higher
Studies (IIHS), Ghaziabad, Uttar Pradesh

CONTENTS

DOI: 10.1201/9781003199403-3

3.1   INTRODUCTION

Big data consists of a group of data sets that are very large that it becomes so challenging to process with the help of available database management tools and mechanisms. Businesses, governmental institutions, HCPs (Health Care Providers), and financial and academic institutions are all taking advantage of the competency of Big Data to work on different business anticipations in addition to enhanced experience by many customers. Almost about 90% of the global data has been created over the past two years. This rate is still growing enormously. In addition, we have noticed that big data has been introduced currently in almost every industry and in every field.

According to Gartner's view, big data can be defined as "Big data" is highly volumed, high velocity, and diversified knowledge assets which requires cost-effective, contemporary mode of knowledge processing for improved insight as well as for good decision making".

There are certain basic points about big data technique that are discussed below:

- This technique points out to large collection of data that is expanding exponentially along with time.
- This technique is so comprehensive that it cannot be refined or evaluated with the help of typical data processing mechanisms.
- This technique consists of data mining process, data storage, data examination, data sharing, and data visualization.
- This technique is all-inclusive one including data, data frameworks, inclusive of various tools and methods that can be used to examine and evaluate the data.

3.1.1   COMPARING CONVENTIONAL DATA TECHNIQUE AND BIG DATA TECHNIQUE

Generally, data is a collection of letters, words, numbers, symbols, or images, but with the progression of different multi-tasking technology-based tools and methods, data has become so distinctive in the context of content as well as origin. With a view

TABLE 3.1
Differences Between Conventional Data and Big Data

	Conventional Data Technique	Big Data Technique
Measure of Capacity	MB, GB	PBs, ZBs
Data Generation Estimate	Long Time Period	More Rapid
Types of Data	Structured	Semi-structured, Unstructured
Data Source	Integrated	Many sources and also distributed
Data Storage	RDBMS(Relational Database Management System)	HDFS and No SQL

of this, big data technique has loomed so far distinct from the old conventional data technique (Table 3.1).

3.2 BIG DATA TECHNIQUE TYPES

Big Data can be classified into three forms:

1. Structured Big Data Type
2. Unstructured Big Data Type
3. Semi-Structured Big Data Type

3.2.1 STRUCTURED BIG DATA TYPE

Structured data is the one which is easily stored, used, and even treated in a steady format. However, they have forecast challenges whenever the size of each and every data expands to a larger extent; exemplary amount is mostly in the range of many zettabytes. 10^{21} bytes is equal to 1 zettabyte or one billion terabytes that forma zettabyte. For case, the table named employee present in a company's database will be ordered as the details of the employee, positions hold by the employee, their income, etc. will exist in a catalogued fashion.

3.2.2 UNSTRUCTURED BIG DATA TYPE

Unstructured data types have no clear format in storage. This makes it very challenging and even stagnant to examine and even evaluate unstructured type of data. Email service is a good example that can be called as unstructured data.

3.2.3 SEMI-STRUCTURED BIG DATA TYPE

Semi-structured data type relates to the data type having formats the same as both structured and unstructured data. An example of semi-structured data is a data expressed in the form of an XML file.

3.3 ESSENCE OF BIG DATA

3.3.1 VOLUME

Organization collects dataset from different sources such as social media platforms, business transactions, as well as information from sensor or machine data. This large collection of data set is reserved in different data warehouses.

3.3.2 VARIETY

Data are not stored in the form of row and column. Data can be structured or unstructured. Log file and CCTV footage are best examples of unstructured data types. Data such as tables are best examples of structured data such as the bank's transactional data.

3.3.3 VELOCITY

Companies are increasingly demanding very hard on how long the data should turn into analytical results based on which users can make decisions. Thus, it is necessary to ensure the collection, storage, processing, and analysis of data in a relatively short time: from one day to the real-time mode.

3.3.4 VARIABILITY

This refers to the deviation which can be demonstrated by the data at different points, thus curbing the process of being able to hold and manage the data efficiently.

3.3.5 VALUE

It represents the value of big data, i.e. it shows the importance of data after analysis. This is due to the fact that the data on its own is almost worthless. The value lies in careful analysis of the exact data, the information and ideas it provides. The value is the final stage that comes after processing volume, velocity, variety, and variability (Figure 3.1).

We are collecting large amounts of information through various types of technologies like computers and satellites. The database management system has stored large amount of data for potent as well as efficient extraction of required information from a large set of data. We have many of the information which can be handled from business transactions data as well as scientific data to get pictures of satellite pictures, text generated reports, and even military intelligence. The huge amount of data set has been reserved in different files, databases, and even other repositories; it is progressively crucial to even construct powerful means for evaluating and perhaps understanding such data set and even for the extirpation of quite compelling knowledge that could be beneficial during decision-creation phase.

Data mining [2] process is also called Knowledge Discovery in Databases (KDD) [1] which means the nontrivial [1] extirpation of implicit, previously unknown

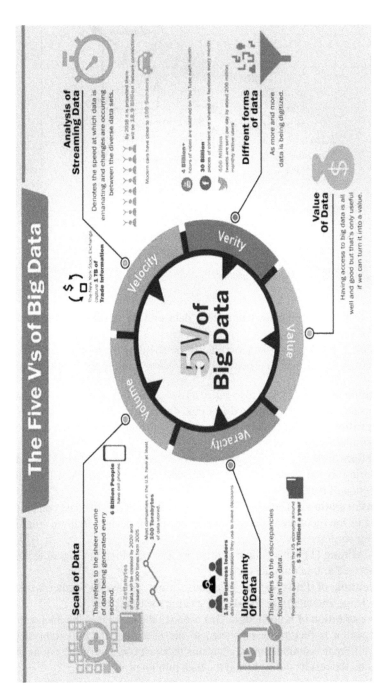

FIGURE 3.1 Five V's of big data.

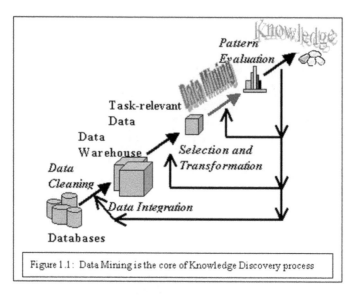

FIGURE 3.2   Knowledge discovery process.

and likely very beneficial knowledge from data present in different databases [1] (Figure 3.2).

The KDD [1] phase consists of phases outstanding from raw form of data collections to new information. The repetitive process contains the following phases:

- **Cleaning of Data [3]**: It is also called as data cleansing in which noise data and inconsistent data are eliminated from the database.
- **Integration of Data**: Data integration is done to combine multiple data sources in a similar source.
- **Selection of Data**: This stage helps to retrieve the data which is relevant according to the evaluation that is decided and extracted from the different collection of data.
- **Transformation of Data [3]**: This phase is also called as the data consolidation phase as the input data is converted into different designs that are suitable for mining purposes.
- **Data Mining [1]**: This is an essential process by which intelligent techniques are applied to find out data patterns.
- **Evaluation of Patterns**: In this phase, truly remarkable patterns defining knowledge are determined based on specific measures.
- **Representation of Knowledge**: This is one of the last stages in which extracted information is pictorially presented to the user. It is an important step that uses different visualization mechanisms to assist users to figure out as well as evaluate the results derived from the data mining process.

The KDD [1] is a progressive phase. When the extracted information is conferred to the end-user as well as the interpretation measures and mechanisms that can be improved, the extraction can be easily refined further. The new information can be chosen or even integrated so as to get different and even more accurate results.

Data mining [1] technique should not be confined only to a single type of data. Data mining [1] should be suitable to other types of stored information. So different algorithms and even different approaches may distinct to many types of data. This technique is being used by various types of databases like relational databases [2], object-relational databases [2] and object-oriented databases [3], data warehouses [1], transactional databases [1], unstructured and even semi-structured [1] repositories such as the World Wide Web(WWW). Data mining is used in advanced databases like spatial databases [1], multimedia databases [1], time-series databases [1] and textual databases [2], and even flat files [2].

3.4 CATEGORIZATION OF DATA MINING SYSTEMS

Data mining [1] systems can be classified on the basis of different measures from other classification [3] techniques as follows:

3.4.1 CLASSIFICATION ON THE BASIS OF THE TYPE OF DATA SOURCE THAT IS MINED

A data mining [3] system can be categorized on the basis of various kinds of databases that are mined such as data models [2] or the types of data [2] or applications [2] that are convoluted. Each database may need its own set of data mining technique.

3.4.2 CLASSIFICATION [3] ON THE BASIS OF KING OF KNOWLEDGE DISCOVERED

Data mining [3] systems can be classified into different kinds of information so that they can be mined easily. It is stationed on the basis of data mining [1] capabilities like characterization [3], discrimination [1], association [1] and correlation analysis [1], classification [1], prediction, clustering [3], outlier analysis, and evolution analysis [3]. An extensive data mining model provides different integrated data mining [1] capabilities.

3.4.3 CLASSIFICATION ON THE BASIS OF THE DATA MODEL ON WHICH IT IS DRAWN

This type of classification categorizes into a data model [3] convoluted like relational database [1], object-oriented database [1], data warehouse [1], transactional database [1].

3.4.4 CLASSIFICATION ACCORDING TO DIFFERENT MINING TECHNIQUES THAT ARE USED

Data mining-based systems contain different techniques and approaches. This type of classification methods can be differentiated into different data analysis methods like

machine learning [1], neural networks [2], genetic algorithms [2], statistics [1], visualization [1], database-oriented [1] or data warehouse-oriented [2].

3.5 DATA MINING DESIGN

The major components that are a part of any data mining system can be classified as
follows.

3.5.1 DATA SOURCE

A huge variety of documents like database [2], data warehouse [1], WWW [1], and
text files [1] becomes substantive data origin. For successful data mining [2] process, a huge collection of factual data is required. The data first should be dredged,
combined, and even preferred before storing it to the database [1] or data warehouse
[2] server. The data comes from many sources and even in different formats. This
cannot be used precisely for the data mining [2] process because there would be
incomplete and unreliable data [2]. So, data is to be cleaned and combined first and
then other data will be gathered from other data sources and only the required data
is selected and moved to the server. These processes [1] are very complex than the
earlier one. A number of approaches may be used on the data as part of the cleaning
phase, integration phase, and even selection phase.

3.5.2 DATA WAREHOUSE [2] SERVER

The database server [2] is the substantial area where data is stored when it is extracted
from [2] different data origins. The server [2] consists of the original amount of data
which is available to be evaluated and therefore the server manages to retrieve the
data. All these processes are performed based on the request of a specific person for
data mining [2].

3.5.3 DATA MINING ENGINE

During the data mining [2] process, the engine [1] becomes the crucial element and
is the used element or a compelling force [2] that holds all the requests [2] made by
the user and even manages [1] them and even consists of a number of modules [1].
The number of modules [1] present in an element includes mining tasks such as classification technique [2], association technique [2], regression technique [2], characterization [1], prediction [1] and clustering [1], time series analysis [2], naive Bayes
[1], support vector machines [2], ensemble methods [1], boosting [1] and bagging [1]
techniques, random forests, and decision trees.

3.5.4 PATTERN ASSESSMENT MODULE

This technique of the components is culpable for gauging the interestingness of
different patterns [1] which can be used for finding out the basic level [1] of the

threshold value and can be used to connect with the data mining engine [1] to corporate in the assessment method of other components. All in all, the main purpose of this element is to find out all the alluring and usable patterns [2] which could contribute in improving software quality.

3.5.5   GRAPHICAL USER INTERFACE (GUI) [2]

When the selected data is being communicated with the engines [1] and between different pattern assessments of different components, it is very necessary to communicate with the different modules present and so it becomes more user friendly so that the effective use of all the present modules could be done and therefore there is the urgency of a graphical user interface commonly known as GUI. It is used to institute a sense of communication between the end-user and the data mining [2] process that, in turn, helps end-users to easily access [3] and even use the system [2] effectively as well as smoothly to keep them barren of any intricacy which has been increasing in the process [2]. This is a type of abstraction [2] in which only the needed modules are displayed to the end-users and all the complexities [1] as well as functionalities [1] that are accountable to construct the system which are hidden for the simplicity purpose. Whenever a query is submitted by the user, the module then easily collaborates with the complete set of a data mining [2] process to produce a needed output which could be easily available to the user in a much more comprehensive manner.

3.5.6   KNOWLEDGE BASE

The knowledge base is beneficial for the data mining stage. It might be meaningful for guiding the search process or for the assessment of the interestingness of the patterns for particular results. This knowledge base [2] comprises different beliefs of users and also the data that is collected from different user experiences which are beneficial for the data mining [2] process. The engine receives its input set from the already available knowledge base, thereby furnishing more effective [1], accurate [1], and even more reliable results. Every module of the data mining technique and even architecture has its own way of achieving the responsibilities (Figure 3.3).

3.6   DATA MINING ARCHITECTURE

3.6.1   ISSUES AND DILEMMA OF BIG DATA TECHNIQUE

3.6.1.1   Dilemma

The main complications in big data technique have grown drastically. The huge collection of data [2] is difficult for the software tools and techniques to manage. Exploring a huge set of data, extracting relevant information from the available data sets and knowledge is a challenge [2]; sometimes it is also a major obstacle. The big data technique [3] is unstructured [3], huge size, and not easy to handle [2].

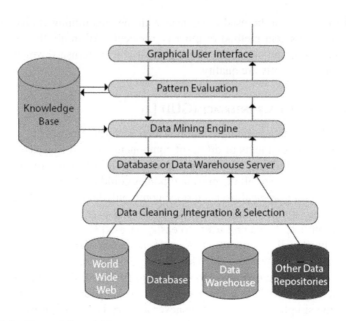

FIGURE 3.3   Data mining architecture.

3.6.1.2   Issues [1]

The main issues [3] of the big data technique are as follows:

a.   Low quality [1] of data such as waste data [1], dirty data [1], and incompetent data size [2].
b.   Duplicated data is transferred from different sources such as multimedia files.
c.   Security [2] issue and even privacy [3] issues related to the companies.
d.   Data mining algorithm [1] is not much effective.
e.   Difficult to process an unstructured data into structured data.
f.   Very costly and less flexible.

3.6.1.3   Solution of Big Data Technique

Hadoop [2] is an open source framework [2] that can be used for keeping and even processing big data [2] in a very dispersed way. It is written in java programming language. It is designed where thousands of machines are scaled up on a single server and each one of them is offering local computation and even storage. The Hadoop Architecture consists of mainly three major layers [2] (Figure 3.4):

•   Processing/Computation layer (MapReduce) [2], and
•   Storage layer (Hadoop Distributed File System) [2].

3.6.1.4   Hadoop Distributed File System

The Hadoop Distributed File System [2] (HDFS) is similar to the Google File System [2] (GFS) that contains a distributed file system that is to be designed in order to

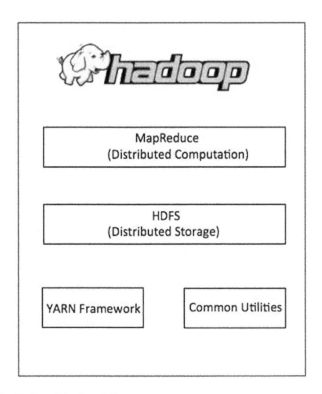

FIGURE 3.4 Hadoop Distributed System.

run on commodity hardware as well. The HDFS provides data storage to Hadoop. It divides the data into blocks and stores them in a distributed manner. The HDFS uses a master-slave architecture.

3.6.1.5 MapReduce

MapReduce [2] is part of a parallel programming model [3] used to write various dispersed applications developed at Google for the effective processing of huge sets of data (multi-terabyte [1] datasets), on larger clusters [2] (thousands of nodes) of commodity hardware [2] in a reliable [2] and even fault-tolerant way. MapReduce [2] is a computational model [2] and even software framework that is used for writing applications that run on Hadoop [2]. These MapReduce [2] programs are useful for evaluating immense data in parallel [2] on large clusters containing computational nodes (Figure 3.5).

3.7 VARIOUS DATA MINING TECHNIQUES TO IMPROVE DATA QUALITY

There are different techniques that help extract data [2], and each and every approach will amass many results. The objective of data mining is to search information that can

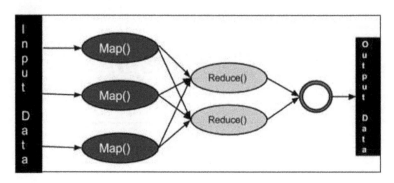

FIGURE 3.5 MapReduce architecture.

be easily understood and can help to improve or enhance data quality management. Each data management software [2] should be able to provide [2] that knowledge from which the end-user can easily gain. Larger data sets [2] are difficult to work on manually. The various techniques to improve data quality are mentioned below.

3.7.1 ANOMALY DETECTION

Looking for knowledge that will not match with anticipated behavior or even a projected pattern [2] is called anomaly detection [2] (or outliers [2], surprises [2], or exceptions [2]). Abnormalities can give litigable information because they diverge from the average [2] value in the selected input set. An outlier [2] is numerically far from other data that is already part of a given input set, and indicates that additional examination is needed for improvement in data quality [2].

3.7.2 CLUSTERING

Clustering [3] is a technique that contributes to find out the same input sets and allows us [2] to know both the similarities [2] and differences [2]of these input sets. Input sets having common properties are used to increase the conversion rate [1]. For example, if the buying trait of one [2] group of customers is the same as that of another group of customers, both could be addressed with the same services or even products [2].

3.7.3 CLASSIFICATION

This approach is used for collecting knowledge about the dataset so that input sets can be classified into proper categories. Email can be further classified into regular [2], acceptable [2] email, or as spam [2]. All of the datasets that are attached to the email are first examined and then categorized as spam [2] if some sender names [2], IP addresses [2], or words [2] are found in the data set.

3.7.4 REGRESSION

Regression analysis [3] is one of the most advanced data mining [2] approaches. It tries to analyze the connectivity between different data elements and represents which variables are easily modified by other variables. Unlike correlation [2], however, regression [2] only represents those variables that are easily modified by others. This approach is used to find out different customer satisfaction [2] levels and how [2] – and to what degree – they modify customer loyalty [2].

3.8 CONCLUSION

Big data [1] is about large volumes of complex data sets. In the fast pace of networking world, storage of data [2], and even data gathering capacity [2], big data is now popularizing in all sciences as well as engineering fields. Data mining has been applied to many fields such as business intelligence [1], Web search [1], scientific discovery [1], and digital libraries [1]. However, big data [2] is facing many problems, dilemmas, and even provide solutions to handle this huge amount of data [2].

This chapter shows the ability of data mining approaches that provide efficient enhancing tools for improving performance in organizations. A comparative analysis of different data mining approaches is presented in this chapter. Many data mining techniques can be easily implemented on datasets to know about their future performance.

REFERENCES

[1] Big Data and Big Data Mining: Study of Approaches, Issues and Future scope International Journal of Engineering Trends and Technology Dec 2014.
[2] http://research.ijcaonline.org/volume74/number5/pxc3889673.pdf.
[3] Xingquan Zhu, Ian Davidson, "Knowledge Discovery and Data Mining: Challenges and Realities", ISBN 978-1-59904-252, Hershey, New York, 2007.
[4] Bisht B., Gandhi P. (2019) "Review Study on Software Defect Prediction Models premised upon Various Data Mining Approaches", INDIACom-2019 10th INDIACom 6th International Conference on "Computing For Sustainable Global Development" at Bharti Vidyapeeth's Institute of Computer Applications and Management (BVICAM).
[5] Gandhi P., Pruthi J. (2020) Data Visualization Techniques: Traditional Data to Big Data. In: Data Visualization. Springer, Singapore. pp. 53–74.

4 Significance of Data Mining in the Domain of Intrusion Detection

Parul Gandhi[1], Ravi Kumar Sharma[2], Tejinder Pal Singh Brar[3], and Pradeep Bhatia[4]
[1]Professor, Faculty of Computer Applications, MRIIRS, Faridabad
[2]Assistant Professor, Department of Computer Applications, CGC Landran, Punjab
[3]Professor and HOD, Department of Computer Applications, CGC Landran, Punjab
[4]Professor, Department of Computer Applications, Guru Jambheshwar University, Hisar

CONTENTS

DOI: 10.1201/9781003199403-4

4.1 INTRODUCTION

Interruption detection is the way toward distinguishing and reacting to vindictive
action focused at registering and systems administration assets. It is a gadget, ordinarily another PC that screens exercises to recognize malevolent or dubious occasions.
An intrusion detection system (IDS) gets crude contribution from sensors, examines
those data sources and afterward makes some move. Since the expense of data handling and Internet openness is dropping, an ever increasing number of associations are
getting defenseless against a wide assortment of digital dangers. As indicated by a
new review by CERT, the pace of digital assaults has been multiplying each year as
of late. Hence, it has gotten progressively imperative to make our data frameworks,
particularly those utilized for basic capacities, for example, military and business
reasons, impervious to and lenient toward such assaults. IDS are an essential piece of
any security bundle of a cutting edge organized data framework. An IDS recognizes
interruptions by observing an organization or framework and investigating a review
stream gathered from the organization or framework to search for hints of pernicious
conduct.

4.2 CLASSIFICATION OF INTRUSION DETECTION SYSTEMS

Intrusion detection systems can be described in terms of three functional components:

1. Information sources: The different sources of data that are used to determine
    the occurrence of an intrusion. The common sources are network, host and
    application monitoring.
2. Analysis: This part deals with techniques that the system uses to detect an
    intrusion. The most common approaches are misuse detection and anomaly
    detection.

3. Response: This implies the set of actions that the system takes after it has detected an intrusion. The set of actions can be grouped into active and passive actions. An active action involves an automated intervention, whereas a passive action involves reporting IDS alerts to humans. The humans are, in turn, expected to take action.

4. Information Sources: Some IDS analyze network packets captured from network bones or LAN segments to find attackers. Other IDS analyze information generated by operating system or application software for signs of intrusion.

4.2.1 NETWORK-BASED IDS

An organization-based IDS dissects network bundles that are caught in an organization. This includes putting a bunch of traffic sensors inside the organization. The sensors ordinarily perform nearby examination and identification and report dubious occasions to a focal area. Most of business interruption discovery frameworks are network based. One benefit of an organization-based IDS is that a couple of very much positioned network-based IDS can screen a huge organization. A detriment of an organization-based IDS is that it can't dissect encoded data. Likewise, most organization-based IDS can't tell if an assault was fruitful; they can just identify that an assault was begun.

4.2.2 HOST-BASED IDS

A host-based IDS dissects review sources, for example, working framework review trails, framework logs or application logs. Since IDS-based frameworks straightforwardly screen the host information records and working framework measures, they can decide precisely which have assets are focuses of a specific assault. Because of the fast advancement of PC organizations, conventional single host interruption location frameworks have been changed to screen various hosts on an organization. They move the checked data from different observed hosts to a focal site for handling. These are named as disseminated interruption identification frameworks. One benefit of a host-based IDS is that it can "notice" the result of an endeavored assault, as it can straightforwardly access and screen the information documents and framework measures that are typically focused by assaults. A detriment of a host-based IDS is that it is more earnestly to oversee and it is more powerless against assaults.

4.2.3 APPLICATION-BASED IDS

Application-based IDS are a unique subset of host-based IDS that dissect the occasion that happen inside a product application. The application log records are utilized to notice the occasions. One benefit of utilization-based IDS is that they can straightforwardly screen the communication between a client and an application which permits them to follow singular clients.

4.2.4   IDS ANALYSIS

There are two primary approaches to analyze events to detect attacks: misuse detection and anomaly detection. Misuse detection is used by most commercial IDS and the analysis targets something that is known to be bad. Anomaly detection is one in which the analysis looks for abnormal forms of activity. It is a subject of great deal of research and is used by a limited number of IDS.

4.2.5   MISUSE DETECTION

This strategy discovers interruptions by checking network traffic looking for direct matches to known examples of assault (called marks or rules). This sort of location is likewise at times called "signature based recognition". A typical type of abuse recognition that is utilized in business items indicates each example of occasions that relates to an assault as a different mark. Notwithstanding, there are more complex methodologies called state-based examination that can use a solitary mark to distinguish a gathering of assaults. An impediment of this methodology is that it can just distinguish interruptions that match a pre-characterized rule. The arrangement of marks should be continually refreshed to stay aware of the new assaults. One benefit of these frameworks is that they have low bogus caution rates. Oddity Detection: In this methodology, the framework characterizes the normal conduct of the organization ahead of time. The profile of ordinary conduct is constructed utilizing strategies that incorporate measurable techniques, affiliation rules and neural organizations. Any critical deviations from this normal conduct are accounted for as potential assaults. The actions and procedures utilized in abnormality recognition incorporate:

- Threshold detection: In this kind of IDS, certain attributes of user behavior are expressed in terms of counts, with some level established as permissible. Some examples of these attributes include the number of files accessed by a user in a given period, the number of failed attempts to login to the system and the amount of CPU utilized by a process.
- Statistical measures: In this case, the distribution of profiled attributes is assumed to fit a pattern.
- Other techniques: These include data mining, neural networks, genetic algorithms and immune system models.

On a fundamental level, the essential benefit of abnormality-based identification is the capacity to distinguish novel assaults for which marks have not been characterized at this point. Be that as it may, by and by, this is hard to accomplish on the grounds that it is difficult to get exact and extensive profiles of typical conduct. This makes an irregularity recognition framework create an excessive number of bogus alerts and it very well may be exceptionally tedious and work serious to filter through this information.

4.2.6   RESPONSE OPTIONS FOR IDS

After an IDS has recognized an assault, it creates reactions. Business IDS uphold a wide scope of reaction choices, arranged as dynamic reactions, inactive reactions or a combination of two. Dynamic responses: Active reactions are mechanized activities taken when particular kinds of interruptions are distinguished. There are three classifications of dynamic reactions.

4.2.7 COLLECT ADDITIONAL INFORMATION

The most well-known reaction to an assault is to gather extra data about a speculated assault. This may include expanding the degree of affectability of data hotspots for instance turn up the quantity of occasions logged by a working framework review trail or increment the affectability of an organization screen to catch every one of the bundles. The extra data gathered can help in settling and diagnosing if an assault is occurring. Another sort of dynamic reaction is to end an assault in progress and square ensuing access by the assailant. Regularly, an IDS achieves this by impeding the IP address from which the assailant seems, by all accounts, to be coming.

4.2.8 TAKE ACTION AGAINST THE INTRUDER

Some folks in the information warfare area believe that the first action in the active response area is to take action against the intruder. The most aggressive form of this response is to launch an attack against the attacker's host or site.

4.2.9 PASSIVE RESPONSES

Passive IDS responses provide information to system users and they assume that human users will take subsequent action based on that information. Alarms and notifications are generated by an IDS to inform users when an attack is detected. The most common form of an alarm is an on screen alert or a popup window. Some commercial IDS are designed to generate alerts and report them to a network management system using SNMP traps. They are then displayed to the user via network management consoles.

4.3 INTRUSION DETECTION ARCHITECTURE

Figure 4.1 shows the different components of the IDS. They are briefly described below.

- Target system: The system that is being analyzed for intrusion detection is considered as the target system. Some examples of target systems are corporate intranets and servers. Feed: A feed is an abstract notion of information from the target system to the intrusion detection system. Some examples of a feed are system log files on a host machine or network traffic and connections.
- Processing: Processing is the execution of algorithms designed to detect malicious activity on some target system. These algorithms can either use signature or some other heuristic techniques to detect malicious activities. The physical architecture of the machine should have enough CPU power and memory to execute the different algorithms.
- Knowledge base: In an intrusion detection system, knowledge bases are used to store information about attacks as signatures, user and system behavior as

profiles. These knowledge bases are defined with appropriate protection and
capacity to support intrusion detection in real time.

- Storage: The type of information that must be stored in an intrusion detection
  system will vary from short-term cached information about an ongoing session
  to long-term event-related information for trend analysis. The storage capacity
  requirements will grow as networks start working at higher speeds.
- Alarms/directives: The most common response of an intrusion detection system
  is to send alarms to a human analyst who will then analyze it and take proper
  action. However, the future trend is for IDS to take some action (e.g. update
  the access control list of a router) to prevent further damage. As this trend continues, we believe that intrusion detection will require messaging architectures
  for transmitting information between components. Such messaging is a major
  element of the Common Intrusion Detection Framework.
- GUI/operator interface: Proper display of alarms from an IDS is usually done
  using a Graphical User Interface. Most commercial IDS have a fancy GUI with
  capabilities for data visualization and writing reports.
- Communications infrastructure: Different components of an IDS and different
  IDS communicate using messages. This infrastructure also requires protection
  such as encryption and access control.

4.4 IDS PRODUCTS

This section presents some of the research and commercial IDS products.

4.4.1 RESEARCH PRODUCTS

4.4.1.1 Emerald

Event Monitoring Enabling Responses to Anomalous Live Disturbances is a research
tool developed by SRI International. They have explored issues in intrusion detection
associated with deviations both from normal user behavior (anomalies) and known
intrusion patterns (signatures).

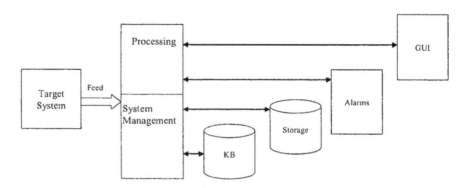

FIGURE 4.1   Intrusion detection architecture.

4.4.1.2 NetStat

This is a research tool produced by the University of California at Santa Barbara. It explores the use of state-transition analysis to detect real-time intrusions.

4.4.1.3 Bro

Bro is a research tool developed by the Lawrence Livermore National Laboratory. The main design goals of Bro are as follows:

a. Highload monitoring.
b. Real time notification.
c. Separating mechanism from policy.
d. Ability to protect against attacks on the IDS.

4.4.1.4 NetProwler

This is a product from Axent Corporation. It supports both host-based and network-based detection. NetProwler provides signatures for a wide variety of operating system and application attacks. It allows a user to build customized signature profiles using a signature definition wizard. Examples of attack signatures that NetProwler supports include the denial of service, unauthorized access, vulnerability probes and suspicious activity that is counter to company policies.

4.4.1.5 NetRanger

This is a product from Cisco Systems. It operates in real time and is scalable to enterprise level. A NetRanger system is composed of Sensors and one or more Directors that are connected by a communication system. In addition to providing many standard attack signatures, NetRanger provides the ability for the user to define their own customized signatures. In response to an attack, the Sensor can be configured with several options that include generating an alarm, logging the alarm event and denying further network access. The Director provides a centralized management support for the NetRanger system. It allows the capability to remotely install new signatures into the Sensors. The Director also provides a centralized collection and analysis of alert data. The status of Sensors can be monitored by the Director using a color-coded scheme.

4.4.2 PUBLIC DOMAIN TOOLS

4.4.2.1 Tripwire

This is a file integrity assessment tool that was originally developed at Purdue University. Tripwire is different from others as it detects changes in the file system of the monitored system. Tripwire comes in both commercial and free versions. Tripwire computes checksums or cryptographic signatures of files. It can be configured to report all changes in the monitored file system. For example, it can check if system binaries have been modified, if syslog files have shrunk or if security settings have unexpectedly changed. It can be configured to perform integrity checks at regular intervals and to provide information to system administrators to implement recovery if tampering has occurred.

4.4.2.2 SNORT

SNORT is an open source NIDS that uses a combination of rules and preprocessors to analyze traffic. SNORT is easy to configure allowing users to create their own signatures and to alter the base functionality using plugins. SNORT has evolved from a simple network management tool to a world-class enterprise distributed intrusion detection system. SNORT detects suspicious traffic by using signature matching. SNORT signatures are written and released by the SNORT community within hours of the announcement of a new security exposure. It has the largest and most comprehensive collection of attack signatures for any IDS.

SNORT uses output plug-ins to store the output of its detection engine. It's outputting functionality is modular and provides different formats (e.g. XML, Relational Database, Text logfile and so on) to store the output. SNORT also provides a GUI to view the alerts. ACID is a Web application that reads intrusion data stored in a database and presents it in a browser in a human friendly format. ACID includes a charting component that is used to create statistics and graphs.

4.4.2.3 Network Flight Recorder

This is a network-based IDS that was previously available in both a commercial version and a public domain version. NFR includes a complete programming language, called N, designed for packet analysis. Filters are written in this language which is compiled into byte code and interpreted by the execution engine. Programs can be written in N to perform pattern matching. Also, functions are provided to store the alert data into a database and perform backend analysis. Some examples of backend analysis are histogram and list. Histogram provides a facility for capturing data in a multi-dimensional matrix. The system can be programmed to generate alerts based on the counts in different cells. The list functions allow records to be stored in a chronological order to store historical information. NFR also provides query backends that allow you to analyze the data. Query backends have their own CGI interface and they also provide graphical functions for data visualization.

4.4.2.4 Intrusion Detection Government Off-the-Shelf (GOTS) Products

This was supported by the Air Force Information Warfare Center. CIDDS receives near real-time connections data and associated events from Automated Security Incident Measurement (ASIM) Sensor host machines and selected other IDS tools. It stores this data on a local database and allows for detailed correlation and analysis by a human analyst.

4.5 TYPES OF COMPUTER ATTACKS COMMONLY DETECTED BY THE IDS

Three types of computer attacks are commonly detected by the IDS: system scanning, denial of service (DOS) and system penetration. These attacks can be launched locally, on the attacked machine or remotely using a network to access the target.

4.5.1 SCANNING ATTACKS

A scanning attack occurs when an attacker probes a target network or system by sending different kinds of packets. From the responses received, the attacker can learn many of the system characteristics and vulnerabilities. Some of the information that the responses can provide are Topology of the target network, Active hosts and operating systems on those hosts. Various tools that are used to perform these activities are network mappers, port mappers, port scanners and vulnerability scanning tools.

4.5.2 DENIAL OF SERVICE (DOS) ATTACKS

Attacks attempt to slow or shut down targeted network systems or services. There are two main types of DOS attacks: flaw exploitation and flooding. Flaw Exploitation DOS Attacks: In this type of an attack, the attacker exploits a flaw in the target system's software. An example of such a processing failure is the "ping of death" attack.

4.5.3 PENETRATION ATTACKS

These attacks involve unauthorized acquisition or alteration of a system resource. Consider these as integrity and control violations. Some examples of penetration attacks are User to Root: A local user on a host gains root access. Remote to User: An attacker on a network gains access to a user account on the target host. Remote to Root: An attacker on the network gains complete control of the target host.

4.6 SIGNIFICANT GAPS AND FUTURE DIRECTIONS FOR IDS

This section discusses significant gaps in the current IDS products.

a. Historical data analysis: As networks are getting large and complex, security officers that are responsible for managing these networks need tools that help in historical data analysis, generating reports and doing trend analysis on alerts that were generated in the past. Current IDS often generate too many false alerts due to their simplistic analysis. The storage management of alerts from IDS for a complex network is a challenging task.

b. Support for real-time alert correlation: Intrusion correlation refers to interpretation, combination and analysis of information from several sensors. For large networks, sensors will be distributed and they will send their alerts to one central place for correlation processing. There is a need for this information to be stored and organized efficiently at the correlation center. Also, traditional IDS focus on low-level alerts and they do not group them even if there is a logical connection among them. As a result, it becomes difficult for human users to understand these alerts and take appropriate actions. It has been reported that for a typical network "users are encountering 10 to 20,000 alerts per sensor per day". Therefore, there is a need to store these alerts efficiently and group them to construct attack scenarios.

c. Heterogeneous data support: In a typical network environment, there are multiple audit streams from diverse cyber sensors: (1) raw network traffic data, (2) netflow data, (3) system calls, (4) output alerts from an IDS and so on. It is important to have an architecture that can integrate these data sources into a unified framework, so that an analyst can access it in real time. Since current IDS are not perfect they produce a lot of false alarms. There is a need for efficient querying techniques for a user to verify if an alert is genuine by correlating it with the input audit data.

d. Forensic analysis: With the rapidly growing theft and unauthorized destruction of computer-based information, the frequency of prosecution is rising. To support prosecution, electronic data must be captured and stored in such a way that it provides legally acceptable evidence.

e. Feature extraction from network traffic data and audit trails: For each type of data that needs to be examined (network packets, host event logs, process traces, etc.) data preparation and feature extraction is currently a challenging task. Due to large amounts of data that needs to be prepared for the operation of IDS, this becomes expensive and time-consuming.

f. Data visualization: During attack, there is a need for the system administrator to graphically visualize the alerts and respond to them. There is also a need to filter and view alerts, sorted according to priority, sub-net or time dimensions. In the next chapter, we will describe how data warehousing and data mining techniques can solve some of these problems in IDS applications.

4.7   DATA MINING FOR INTRUSION DETECTION

Recently, there is a great interest in application of data mining techniques to intrusion detection systems. The problem of intrusion detection can be reduced to a data mining task of classifying data. Briefly, one is given a set of data points belonging to different classes (normal activity, different attacks) and aims to separate them as accurately as possible by means of a model. This section gives a summary of the current research project in this area.

4.7.1   ADAM

The ADAM project at George Mason University [1], [2] is a network-based anomaly detection system. ADAM learns normal network behavior from attack-free training data and represents it as a set of association rules, the so-called profile. At run time, the connection records of past delta seconds are continuously mined for new association rules that are not contained in the profile. ADAM is an anomaly detection system. It is composed of three modules: a preprocessing engine, a mining engine and a classification engine. The preprocessing engine sniffs TCP/IP traffic data and extracts information from the header of each connection according to a predefined schema. The mining engine applies mining association rules to the connection records. It works in two modes: training mode and detecting mode. In the training mode, the mining engine builds a profile of the users and systems normal behavior and generates association rules that are used to train the classification engine. In the detecting mode, the mining engine mines unexpected association rules that are different from the profile. The classification engine will classify the unexpected association rules into normal and abnormal events. Some abnormal events can be further classified as attacks. Although the mining of association rules has used previously to detect intrusions in audit trail data, the ADAM system is unique in the following ways:

- It is online; it uses an incremental mining (online mining) which does not look at a batch of TCP connections, but rather uses a sliding window of time to find the suspicious rules within that window.
- It is an anomaly detection system that aims to categorize using data mining the rules that govern the misuse of a system. For this, the technique builds, apriori, a profile of "normal" rules, obtained by mining past periods of time in which there were no attacks. Any rule discovered during the online mining that also belongs to this profile is ignored, assuming that it corresponds to a normal behavior.

ADAM performs its task in two phases. In the training phase, ADAM uses a data stream for which it knows where the attacks are located. The attack-free parts of the stream are fed into a module that performs offline association rules discovery. The output of this module is a profile of rules that we call "normal", i.e. it provides the behavior during periods when there are no attacks. The profile along with the training data set is also fed into a module that uses a combination of dynamic, online algorithm for association rules, whose output consists of frequent item sets that characterize attacks to the system. These item sets are used as a classifier or decision tree. This whole phase takes place offline before we use the system to detect attacks. The second phase of ADAM in which we actually detect attacks. Again, the online association rules mining algorithm is used to process a window of current connections. Suspicious connections are flagged and sent along with their feature vectors to the trained classifier, where they are labeled as attacks, false alarms or unknown. When the classifier labels connections as false alarms, it is filtering them out of the attacks set and avoiding passing these alerts to the security officer. The last class, i.e. unknown is reserved for the events whose exact nature cannot be confirmed by the classifier. These events are also considered as attacks and they are included in the set of alerts that are passed to the security officer.

4.7.2 MADAM ID

The MADAM ID project at Columbia University [3], [4] has shown how data mining techniques can be used to construct an IDS in a more systematic and automated manner. Specifically, the approach used by MADAM ID is to learn classifiers that distinguish between intrusions and normal activities. Unfortunately, classifiers [5] can perform really poorly when they have to rely on attributes that are not predictive of the target concept. Therefore, MADAM ID proposes association rules and frequent episode rules as means to construct additional more predictive attributes. These attributes are termed disfeatures.

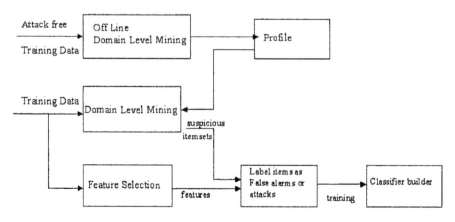

FIGURE 4.2 The training phase of ADAM.

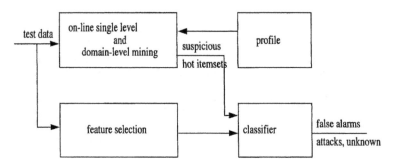

FIGURE 4.3 The intrusion detection phase of ADAM.

We will describe briefly how MADAM ID is used to construct network-based misuse detection systems. First all network traffic is preprocessed to create connection records. The attributes of connection records are intrinsic connection characteristics such as source host, the destination host, the source and destination posts, the start time, the duration, header flags and so on. In the case of TCP/IP networks, connection records summarize TCP sessions.

The most important characteristic of MADAM ID is that it learns a misuse detection model from examples. In order to use MADAM ID, one needs a large set of connection records that have already been classified into "normal records" or some kind of attacks. MADAM ID proceeds in two steps. In the first step it does feature construction in which some additional features are constructed that are considered useful for doing the analysis. One example for this step is to calculate the count of the number of connections that have been initiated during the last two seconds to the same destination host as the current host. The feature construction step is followed by the classifier learning step. It consists of the following process:

• The training connection records are partitioned into two sets, namely normal connection records and intrusion connection records.

- Association rules and frequent episode rules are mined separately from the normal connection records and from the intrusion connection records. The resulting patterns are compared and all patterns that are exclusively contained in the intrusion connection records are collected to form the intrusion only patterns.
- The intrusion only patterns are used to derive additional attributes such as count or percentage of connection records that share some attribute values with the current connection records.
- A classifier is learned that distinguishes normal connection records from intrusion connection records; this classifier is the end product of MADAM ID.

4.7.3 MINDS

The MINDS project [6] [7] at the University of Minnesota uses a suite of data mining techniques to automatically detect attacks against computer networks and systems. Their system uses an anomaly detection technique to assign a score to each connection to determine how anomalous the connection is compared to normal network traffic. Their experiments have shown that anomaly detection algorithms can be successful in detecting numerous novel intrusions that could not be identified using widely popular tools such as SNORT. Input to MINDS is Netflow data that is collected using Netflow tools. The netflow data contains packet header information, i.e. they do not capture message contents. Netflow data for each 10 minute window which typically results in 1 to 2 million records is stored in a flat file. The analyst uses MINDS to analyze these 10 minute data files in a batch mode. The first step in MINDS involves constructing features that are used in the data mining analysis. Basic features include source IP address and port, destination IP address and port, protocol, flags, number of bytes and number of packets.

Derived features include time-window and connection-window-based features. After the feature construction step, the data is fed into the MINDS anomaly detection module that uses an outlier detection algorithm to assign an anomaly score to each network connection. A human analyst then has to look at only the most anomalous connections to determine if they are actual attacks or other interesting behavior. MINDS uses a density-based outlier detection scheme for anomaly detection. The reader is referred to [6] for a more detailed overview of their research. MINDS assigns a degree of being an outlier to each data point which is called the local outlier factor (LOF). The output of the anomaly detector contains the original Netflow data with the addition of the anomaly score and relative contribution of the different attributes to that score. The analyst typically looks at only the top few connections that have the highest anomaly scores. The researchers of MINDS have [8] their system to analyze the University of Minnesota network traffic. They have been successful in detecting scanning activities, worms and non-standard behaviors such as policy violations and insider attacks.

4.7.4 CLUSTERING OF UNLABELED ID

Traditional anomaly detection systems require "clean" training data in order to learn the model of normal behavior. A major drawback of these systems is that clean training data is not easily available. To overcome this weakness, recent research has investigated the possibility of training anomaly detection systems over noisy data [9]. Anomaly detection over noisy data makes two key assumptions about the training data. First, the number of normal elements in the training data is assumed to be significantly larger than the number of anomalous elements. Second, anomalous elements are assumed to be qualitatively different from normal ones. Then, given that anomalies are both rare and different, they are expected to appear as outliers that stand out from the normal baseline data. Portnoy et al. [9] apply clustering to the training data. Here the hope is that intrusive elements will bundle with other intrusive elements, whereas normal elements will bundle with other normal ones. Moreover, as intrusive elements are assumed to be rare, they should end up in small clusters. Thus, all small clusters are assumed to contain intrusions/anomalies, whereas large clusters are assumed to represent normal activities. At run time, new elements are compared against all clusters and the most similar cluster determines the new element's classification as either "normal" or "intrusive".

4.7.5 ALERT CORRELATION

Correlation techniques from multiple sensors for large networks is described in [10] and [11]. A language for modeling alert correlation is described in [12]. Traditional IDS focus on low-level alerts and they raise alerts independently though there may be a logical connection between them. In case of attacks, the number of alerts that are generated become unmanageable. As a result, it is difficult for human users to understand the alerts and take appropriate actions. Ning et al. present a practical method for constructing attack scenarios through alert correlation, using prerequisites and consequences of intrusions. Their approach is based on the observation that in a series of attacks, alerts are not isolated, but related as different stages, with earlier stages preparing for the later ones. They proposed a formal framework to represent alerts with their prerequisites and consequences using the concept of hyper-alerts. They evaluated their approach using the 2000 DARPA intrusion detection scenario specific datasets.

4.8 CONCLUSIONS

We reviewed the application of data mining techniques to the area of computer security. Data mining is primarily being used to detect intrusions rather than to discover new knowledge about the nature of attacks. Moreover, most research is based on strong assumptions that complicate building of practical applications. First, it is assumed that labeled training data is readily available, and second it is assumed that this data is of high quality. Different authors have remarked that in many cases, it is not easy to obtain labeled data. Even if one could obtain labeled training data by simulating intrusions, there are many problems with this approach. Additionally, attack simulation limits the approach to the set of known attacks. We think that the difficulties associated with the generation of high quality training data will make it difficult to apply data mining techniques that depend on availability of high quality labeled training data. Finally, data mining in intrusion detection focuses on a small subset of possible applications. Interesting future applications of data mining might include the discovery of new attacks, the development of better IDS signatures and the construction of alarm correlation systems.

Finally, data mining projects should focus on the construction of alarm correlation systems. Traditional intrusion detection systems focus on low-level alerts and they raise alerts independently even though there is a logical connection among them. More work needs to be done on alert correlation techniques that can construct "attack strategies" and facilitate intrusion analysis. One way is to store data from multiple sources in a data warehouse and then perform data analysis. Alert correlation techniques will have several advantages. First, it will provide a high-level representation of the alerts along with a temporal relationship of the sequence in which these alerts occurred. Second, it will provide a way to distinguish a true alert from a false alert. We think that true alerts are likely to be correlated with other alerts, whereas false alerts will tend to be random and, therefore, less likely to be related to other alerts. Third, it can be used to anticipate the future steps of an attack and, thereby, come up with a strategy to reduce the damage.

REFERENCES

[1]   Barbara, D., Wu, N., & Jajodia S., Detecting novel network intrusions using Bayes estimators. In Proc. First SIAM Conference on Data Mining, Chicago, IL, April 2001.

[2]   Basheer, S., Nagwanshi, K. K., Bhatia, S., Dubey, S., & Sinha, G. R. FESD: An approach for biometric human footprint matching using fuzzy ensemble learning. IEEE Access, 9, 26641–26663, 2021.

[3]   Lee,W. , Stolfo, S. J., & Kwok, K. W. Mining audit data to build intrusion detection models. In Proc. Fourth International Conference on Knowledge Discovery and Data Mining, New York, 1998.

[4]   Lee, W. & Stolfo, S. J. Data Mining approaches for intrusion detection, In Proc. Seventh USENIX Security Symposium, San Antonio, TX, 1998.

[5]   Han, J. & Kamber, M. Data Mining: Concepts and Techniques, Morgan Kaufmann; 1st edition (6 September 2000), ISBN: 1558604898.

[6]   Ertoz, L., Eilertson, E., Lazarevic, A., Tan, P., Dokes,P., Kumar, V., & Srivastava, J. Detection of Novel Attacks using Data Mining, Proc. IEEE Workshop on Data Mining and Computer Security, November 2003.

[7]   Dahiya, K., & Bhatia, S. Customer churn analysis in telecom industry. In 2015 4th International Conference on Reliability, Infocom Technologies and Optimization (ICRITO) (Trends and Future Directions) (pp. 1–6). IEEE, 2015.

[8]   Bisht, B., Gandhi, P. "Review Study on Software Defect Prediction Models premised upon Various Data Mining Approaches", INDIACom-2019 10th INDIACom 6th International Conference on "Computing For Sustainable Global Development" at Bharti Vidyapeeth's Institute of Computer Applications and Management (BVICAM), 2019.

[9]   Portnoy, L., Eskin, E., & Stolfo, S. J. Intrusion Detection with unlabeled data using clustering. In Proceedings of ACM Workshop on Data Mining Applied to Security, 2001.

[10] Ning, P., Cui, Y., & Reeves, D. S. Constructing Attack Scenarios through Correlation of Intrusion Alerts, Proc. ACM Computer and Communications Security Conf., 2002.

[11] Ning, P. & Xu, D. Earning Attack Strategies from Intrusion Alerts, Proc. ACM Computer and Communications Security Conf., 2003.

[12] Bhatia, S. A Comparative Study of Opinion Summarization Techniques. IEEE Transactions on Computational Social Systems, 8(1), 110–117, 2020.

[13] Gandhi, P. & Pruthi, J. Data Visualization Techniques: Traditional Data to Big Data. In: Data Visualization. Springer, Singapore, pp.53–74, 2020.

5   Data Analytics and Mining

Platforms for Real-Time Applications

Saima Saleem[1] and Monica Mehrotra[2]
[1]Scholar, Department of Computer Science,
Jamia Millia Islamia, New Delhi, India
[2]Professor, Department of Computer Science,
Jamia Millia Islamia, New Delhi, India

CONTENTS

DOI: 10.1201/9781003199403-5

5.1 INTRODUCTION

Due to the internet explosion and fast technological progress, unfathomable amounts of diverse data are emitted at high speeds from numerous sources (social networks, marketing, government, health) which cannot be managed by traditional processing and storage systems such as Relational Database Management Systems (RDBMS). This data is called big data [1], which contrasts from regular data in certain characteristics known as 7v's [2] as illustrated in Figure 5.1. Big data offers incredible opportunities and transformative potential for numerous sectors. But this data isn't valuable all by itself. Rather, the actual value and the benefits lie in the meaningful insights that can be extracted from it and this is where data analytics comes in. Big data analytics is the process of exploring big data using cutting-edge analytic techniques with an objective to infer and draw conclusions about that information. Data mining is a key aspect of analytics. It is an indispensable process in analytics [3] for uncovering interesting and valuable patterns from massive datasets [4]. Traditional data analytics is centered on structured data, i.e. traditional data is organized data with a fixed format. It doesn't cope with the diverse nature of data that organizations are accumulating such as audio, video, and images, which do not have a fixed structure known as unstructured data [5]. The most notable platform for handling big data is Apache Hadoop [6]. It is an open-source framework for big data processing. It distributes storage and processing across a set of interconnected nodes called a cluster. The main component for data processing in the Hadoop framework is MapReduce [7] exclusively developed for batch processing and extremely high-throughput jobs. It efficiently endows valuable insight into what has happened in the past but it is not suitable for various applications that require an instant analysis; therefore, it is incapable of satisfying the constraints in real-time [8]. Due to slower

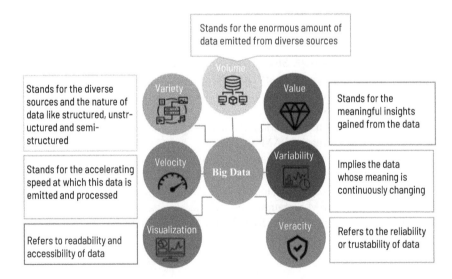

FIGURE 5.1   7v's of big data.Out of 7v's of big data, the value is the final stage that comes after processing volume, velocity, variety, and variability.

response time and high latency, it isn't completely suitable to handle the dynamic real-time data. Hence, advanced solutions and platforms have been introduced for these new demands and we discuss these in this chapter.

The chapter has been structured as follows. Real-time analytics has been explained in Section 5.2, applications of real-time analytics are covered in Section 5.3, a brief introduction to real-time architectures is presented in Section 5.4, and Section 5.5 deals with real-time analytics platforms and is concluded with their comparison.

5.2   REAL-TIME ANALYTICS

Real-time analytics is a sort of big data analytics in which colossal amounts of data are processed and analyzed strictly within a specific timeline. It can deliver critical insights and conclusions can be drawn with very low latency [9]. Real-time applications depend on instantaneous inputs and processes within a certain time span and produce a quick response. By and large, if a decision cannot be made within that time span, it becomes worthless. Therefore, it is critical to process and analyze the data in the right time [10]. Many real-time applications involve an endless flow of unbounded data that is transferred from one location to another over time called stream data, e.g. emitted by sensors, intelligent transportation systems, stock markets, IoT, Recommendation systems, and video surveillance systems. Such kind of data updates with high frequency loses its value and relevance in a brief timeframe. Therefore, it becomes imperative to process and analyze such data on-the-fly enabling time-critical decision making in numerous applications. However, the standard Hadoop-MapReduce is centered on a disk approach, i.e. it reads and writes the data to and from the disk and each iteration outcome gets written to the disk that takes a significant amount of time, thereby making it very slow. It can take minutes to hours to produce an output which makes it inappropriate for such time-sensitive applications [11].

5.2.1   STREAM PROCESSING

Contrary to batch processing, in which the data is bounded with a beginning and an end stored over a period of time on some permanent storage, stream processing a big data technology enables the processing of an infinite stream of data, continuous query, and real-time analytics as the data arrives [11]. The streaming data system expedites the computation speed by utilizing in-memory computing. Thus, it obtains exceptionally low latency and efficiently copes with real-time stream data [12].

5.2.2   IN-MEMORY COMPUTING

Traditional systems have been based on slow disk storage (secondary memory) and relational databases using a Structured Query language (SQL) [13]; all the analytics have been carried out after retaining data on secondary storage that ultimately have extremely high access latency. Dealing with immense volumes of data makes hard disks unseemly for doing real-time data analytics, particularly when the data is largely unstructured [14]. In-memory computing technology utilizes a Random

Access Memory (Main memory) for caching and processing big data across numerous computers that offers very high bandwidth in comparison to disks enabling real-time analysis of enormous amounts of data at very high speeds, therefore reducing the access latency drastically [15].

5.3 APPLICATIONS OF REAL-TIME ANALYTICS

The significance and needfulness of real-time data analytics are in almost every field; in Table 5.1, we highlight a few domains where the real-time data analytics application is extremely helpful and demanding.

5.4 A BRIEF INTRODUCTION TO REAL-TIME BIG DATA ARCHITECTURES

The two most important architectures for real-time big data processing are as follows:

5.4.1 LAMBDA ARCHITECTURE

Lambda Architecture has been developed by Nathan Marz for handling big data by exploiting both batch and stream processing models [24]. It includes three distinct layers: batch layer, speed layer, and serving layer as shown in Figure 5.2. The batch layer generates batch views from the batch data stored in a distributed file system [25] which generally takes very high latency. To complement the batch views, the speed layer is used that generates real-time views from the real-time data stream. Instead of continuously re-computing the batch views by the batch layer, an incremental approach is employed by the speed layer which increments the real-time views based on the latest data [26]. The serving layer performs indexing and amalgamates batch views and real-time views from batch and speed layers, respectively, thus enabling easy access for users [27].

5.4.2 KAPPA ARCHITECTURE

The Kappa Architecture shown in Figure 5.3 was introduced by Jay Kreps to palliate the complexities of Lambda Architecture by eradicating the need for maintaining two distinct sets of code for batch layer and speed layer. It fuses the batch layer and the stream layer together. It performs real-time processing and continuous re-processing of data with the only stream processing engine. The Kappa Architecture comprises a stream processing layer and a serving layer. Contrary to the batch layer, the stream processing layer does not have a beginning or an end from a temporal aspect. It continuously processes new data as it arrives. The serving layer is employed for querying. All the computations take place as stream processing and no batch re-computations are performed on the entire dataset. Re-computations are only performed to support changes and new requirements [24–26].

TABLE 5.1
Applications of Real-Time Analytics

Emergency Response

Sudden onset emergencies such as natural or anthropogenic disasters bring uncertainties and a rising need to provide low latency and precise information, and early warnings to formal response organizations. So, appropriate measures can be taken to prevent the associated risks. An early warning system entails time-critical processing of data collected in real-time from umpteen sources like sensors, unmanned aerial vehicles (UAVs), satellites along with weather data and geographic maps to predict where the natural disaster will occur, e.g. Indian Tsunami Early Warning System for Tsunami Detection [16]. Therefore, real-time analytics can facilitate in preventing loss of human lives and other damages in times of crisis [17].

Transportation

To enable efficient transportation system, real-time analytics is of high demand as massive volumes of data generated from various kinds of sensors like road sensors, vehicle sensors as well as data from satellite require swift processing within a very brief time for various intelligent transportation services. With real-time data analytics, an end-user can make a very smart decision and can react in real-time. Road traffic condition monitoring facilitates us in choosing the best path for the destination; taking a different and safest route in case of unanticipated events such as traffic accidents, adverse weather conditions; dynamically determining time for various emergency rescue vehicles like ambulances, fire fighting service cars for the fast arrival to the destination [18].

Healthcare

Real-time analytics has revolutionized the field of healthcare by improving the condition of healthcare services. Massive healthcare data is emitted from various sources like clinical data, genomic data, real-time stream monitoring from medical devices, and so on. Analyzing this data in an appropriate and reliable way and generating real-time response have improved the quality of clinical care for the patients [19]. Such a statistical method is proposed in [20] that employs real-time analytics to predict the survival rate and length of stay of diabetic patients; a Telehealth system is proposed in [21] that endows real-time information updates on patients' health. It keeps a track of the real-time physiological data using numerous sensors to deal with telemedicine.

Internet of Things (IoT)

IoT is the system of devices, sensors, controllers, and appliances connected to the wider internet that emit continuous massive amounts of data; such data are transient in value; therefore, it requires real-time analysis for effective decision making. For instance, Smart Power Grids monitor the consumers by Smart meters and provide real-time information about the total usage of electricity [22]. Other examples are smart agriculture, smart sports, smart cities, etc. [23].

Stock Market

Although data mining and Business Intelligence techniques are widely employed in the finance domain, these techniques have very high latencies and only work on historical data. Therefore, they are inappropriate for early detection and prevention processes for many financial systems like automated trading and fraud detection systems. With real-time analytics, dealing with such processes can be performed easily and with very low latency. Massive data emitted from stock markets can be analyzed in real-time, which can help in detecting illegitimate activities, timely detection of financial threads, share price forecasting, early decision making for buying or selling shares, etc., thereby boosting the overall performance of the stock market [10, 18].

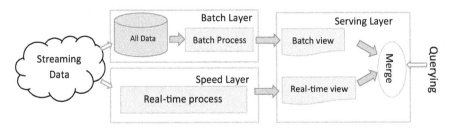

FIGURE 5.2 Lambda architecture.

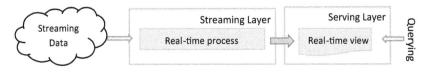

FIGURE 5.3 Kappa architecture.

5.5 REAL-TIME ANALYTICS PLATFORMS

In this section, we discuss open-source as well as commercial real-time analytics platforms.

5.5.1 OPEN-SOURCE PLATFORMS

5.5.1.1 Apache Storm

Apache Storm is the most popular, free, and distributed real-time platform released and open-sourced by Twitter for processing infinite streaming data at a fast speed written in Clojure programming language [28]. Apache Storm makes use of an independent workflow, directed acyclic graphs (DAGs) in its platform [14], and carries out in-memory data processing which makes Storm appropriate for real-time analytics. Thus, it enables us to develop real-time analytics solutions. Storm doesn't come up with its own data storage, but it is capable enough to integrate with other data storage platforms [29]. Storm supports a multitude of programming languages. The key abstractions which perform tasks in Apache Storm illustrated in Figure 5.4 are as follows:

- Topology: It is a Storm program which is represented as a DAG where each node in the DAG is a spout or bolt and each edge in the DAG specifies the flow of tuples [9].
- Stream: It is a boundless series of tuples. A tuple is a fundamental unit of data.
- Spout: A spout is a stream producer. It reads data from various data sources and passes the data to bolt to be processed [13].
- Bolt: All processing is done in bolts; they receive data from spouts, process it, and forward it to the next stage in the topology. Bolts can be exploited to

Topology

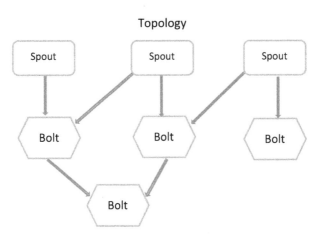

FIGURE 5.4 Storm abstractions.

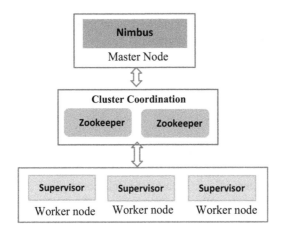

FIGURE 5.5 High-level architecture of storm.

carry out any sort of processing like filtering, aggregations, joining data, and connecting to databases [9].

Storm is based on a master-slave architecture where Nimbus and Supervisors are master and slaves, respectively, as shown in Figure 5.5.

Storm comprises several components:

- Master Node: It runs a master daemon called Nimbus [30]. Topologies are submitted to the master node. Nimbus distributes codes across the cluster, allocates tasks among the worker nodes, and keeps a track of their performance, and reallocates workers as required.
- Processing Nodes/Worker Nodes: The processing node runs worker daemon called 'Supervisor' which governs the worker processes. The real execution

of streaming applications is done in worker nodes. Based on the signals from Nimbus, it can start or stop the worker processes. In case the worker process repeatedly fails on startup and is not able to send any heartbeat messages, it is then rescheduled by Nimbus [31].

- Coordinator (Zookeeper): Apache Storm depends on Zookeeper for cluster-state maintenance. It performs all the coordination between Nimbus and Supervisors [32].

The following are the key features of Apache Storm [2, 31, 33]:

- Fast: Apache Storm is extremely fast as it is able to process a million tuples per second per node.
- Fault-Tolerant: Storm can automatically restart workers when they die. A worker is reassigned to another node if restarting fails repeatedly.
- Scalable: Nodes can be very easily added or removed from the Storm cluster without deranging existing data flows by means of Storm topologies.
- Reliable: Strom ensures that it will always process each tuple at-least-once. Storm has a higher-level API called Trident that enables exactly once messaging semantics for most computations.
- Latency: Because a tuple is processed as soon as it comes up, the latency of Storm is sub-second.
- Apache Storm is embraced by several organizations because of its real-time processing features. Some of the use cases are Twitter, Yahoo, Groupon, Spotify, Yelp, and so on.

5.5.1.2   Apache Spark Streaming

Apache Spark is an open-source and distributed framework for big data processing. Spark performs in-memory computation to enhance the performance of big data analytics applications. It fully supports Lambda Architecture. Spark Streaming enhances the key Spark API allowing scalable, fault-tolerant, exceptionally high-throughput live data stream processing. Data can be consumed from numerous streaming data sources like Kinesis, Kafka, Apache Flume, or Twitter API, and using complex algorithms with various kinds of high-level functions, this data can be processed. And ultimately, the processed data can be sent out to databases such as memsql, Cassandra, and so on, file systems such as HDFS, and live dashboards [34] as shown in Figure 5.6.

FIGURE 5.6   Spark Streaming architecture [44].

FIGURE 5.7 Spark Streaming data flow [44].

Rather than processing the data on-the-fly, Spark Streaming implements a micro-batching approach, i.e., it splits the input data into a series of batches or micro-batches as shown in Figure 5.7. DStream or Discretized stream is the basic abstraction for data streams, which consists of a continuous sequence of resilient distributed datasets (RDD) entailing data of a specific stream interval. Transformations can be applied to these DStreams. Spark core engine processes each batch as RDD and delivers results of RDD operations in batches. Programs for Spark Streaming can be written using Java, Python, or Scala programming languages. Despite Spark being so powerful than Hadoop-MapReduce, Spark is not a true streaming engine as it follows a micro-batching approach [11, 35]. This is on the grounds that the increase in the data input rate results in more and more buffering of data to form a mini-batch in a set time slot, leading to an increase in latency. Typical use cases include website monitoring, fraud detection, and Ad monetization [13, 28, 32].

The following are the key features of Spark Streaming [28]:

- High Throughput: The buffer enables Spark Streaming to cope with the enormous bulk of data, so there is a huge increase in the overall throughput.
- Fault-Tolerant: It recovers quickly from failures. Spark Streaming is capable of recovering both lost work and operator state without any need to write additional code.
- Exactly Once Guarantee: Each message is guaranteed to be processed exactly once.
- Integration: Spark Streaming amalgamates batch and real-time processing enabling us to utilize the same code for batch processing and execute interactive queries on the stream state.

5.5.1.3 Apache Samza

Apache Samza is distributed and open-source stream processing platform written in Scala and Java programming languages [36] centered on the publish/subscribe model. Apache Samza enables real-time processing and analysis of data [37]. It utilizes Apache Kafka for communications and depends on Apache YARN for the distribution of tasks among applications. Kafka provides replicated storage of data that can be accessed with low latency; therefore, Samza jobs can have subsecs latency when executing with it. In Samza, tasks are enabled to maintain state by keeping it on disk (generally with the help of Kafka). To avoid performance issues, this state is kept on the same machine as the processing task which helps Samza in achieving high throughput. It keeps track of the delivery of the message and if there is a failure, it

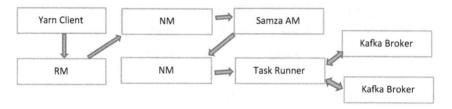

FIGURE 5.8   Apache Samza architecture.

re-sends the message; it employs a check-pointing system to avoid any data loss. Samza offers at-least-once guarantees, i.e. every message in the input stream will be processed at-least-once [28]. One of the main differences between Samza and Storm is that Samza requires YARN and Samza has a very simple and less configurable parallelism model than Storm [38]. Figure 5.8 shows the high-level architecture of Samza.

The core components of Samza are Kafka and YARN. Apache Kafka is a messaging system that enables the collection and delivery of enormous amounts of stream data [32]. Apache YARN is part of the Hadoop project which enables the execution of distributed applications on a cluster. It consists of three components: a Resource Manager (RM), a Node Manager (NM), and an Application Master (AM). The RM manages the resources in the cluster and allocates them to applications. Every node in the cluster has an NM, which is responsible for managing containers on that node – starting them, monitoring their resource utilization, and reporting it to the RM. A coordinator, which is called an AM, is implemented to execute applications on the cluster. The duty of AM is requesting resources, including CPU, memory from the RM on behalf of the application. Samza provides its own implementation of the AM for each job. The Samza AM is the control-hub for a Samza application that runs on a YARN cluster. The Samza client employs YARN for executing a Samza job: YARN then oversees Samza Containers where the processing code runs using the StreamTask API. All the input and output for the Samza StreamTasks originate from Kafka brokers that are generally co-located on the same machines as the YARN NMs [39, 40].

The following are the key features of Apache Samza [41]:

- State Management: Samza handles the snapshotting and restoration of the state of a stream processor.
- Fault-Tolerance: In case a machine in the cluster fails, Samza together with YARN shifts the tasks along with their respective states to other machines.
- Durability: By employing Kafka, Samza ensures that the processing of messages is done in the manner they were written to a partition and prevents the loss of messages.
- Scalability: Samza is horizontally scalable.

FIGURE 5.9   Flink architecture.

- Pluggable: A pluggable API is provided by Samza that enables Samza to integrate with other messaging systems and execution environments like Kafka, AWS Kinesis, Azure Eventhubs, ElasticSearch, and Hadoop.

5.5.1.4   Apache Flink

Apache Flink is an open-source distributed framework with stream and batch processing capabilities written in Scala and Java programming languages [42]. It provides great support for data analytics in real-time. Flink is a realization of the Kappa Architecture. It is based on a streaming execution model but it can process both bounded and unbounded data, with two APIs running on the same distributed streaming execution. Flink is also a powerful tool for batch processing [36]. Flink's core streaming data-flow engine executes data-flow programs [43], and provides communication, data distribution, and fault-tolerance for distributed computations over data streams. Batch processing is built on top of the streaming engine, overlaying native iteration support, managed memory, and program optimization [44]. The arbitrary data stream programs are executed in a data-parallel and pipelined manner, which results in achieving low latency. Flink implements a lightweight fault-tolerance mechanism, which is based on distributed checkpoints. It takes consistent snapshots of the current state of the distributed system at regular intervals without missing any information and without recording duplicates which are then stored to durable storage. If there is any failure, the most recent snapshot is restored, and the stream source is backtracked to the moment that the snapshot was drawn and is replayed. Flink doesn't have its personal storage system; yet, it confers data-source and sink connectors to systems like Kafka Kinesis and Cassandra. Flink has a master-slave architecture. Figure 5.9 shows various components of Flink architecture: Job Manager, Task Manager(s), and clients [28, 32, 42, 43].

- Client: A Flink program is submitted to a client. It converts the Flink program to a DAG and then passes it to the Job Manager.
- Job Manager: It is the master daemon responsible for handling all the computations in the Flink system. It performs the allocation of resources, scheduling of tasks, monitoring the progress and state of each operator, and stream. It distributes the work and submits it to the Task Managers.

- Task Manager: It is the slave daemon. All data processing occurs in Task Managers. Task Managers execute operators that generate data streams and reports to the Job Manager about their status.

Unlike Spark which utilizes batching to imitate streaming computations, Flink is inherently a stream processing engine. Flink is quite similar to Storm as far as the underlying implementation is concerned; however, it is the data abstractions that remain different [35].

The following are the key features of Apache Flink [42, 43]:

- Flink is highly fault tolerant. It attains fault-tolerance by implementing a check-pointing mechanism.
- Flink provides reliable execution with exactly once processing guarantees.
- Flink has a single execution environment for stream processing and batch processing.
- It can process events swiftly with exceptionally low latency (subsecs).

5.5.1.5 Apache Kafka

Apache Kafka is a distributed streaming platform [45] and an open-source message broker project written in Scala and Java. A messaging system enables data exchange between applications, servers, and processes enabling them to focus on the data without worrying about how to share it; this is where Apache Kafka comes in; it processes and mediates communication between applications. Organizations such as NETFLIX, UBER, and Walmart make use of Apache Kafka [46]. Kafka is fundamentally used to develop real-time streaming data pipelines that reliably move data between systems or applications and real-time streaming applications that transform the streams of data. It amalgamates messaging, storage, and stream processing to enable the storage and analysis of both historical and real-time data. The traditional messaging models are queuing model and publish-subscribe model. The queuing allows the distribution of data processing across numerous consumer instances, enabling high scalability. However, queues aren't multi-subscriber, i.e. once one process reads the data, it's gone. Publish-subscribe model allows the broadcasting of data to numerous processes; however, it can't scale processing. Kafka amalgamates these two messaging models and has both of these properties such as it can scale processing and is also multi-subscriber, i.e. there is no compelling reason to pick either [47].

The following are the core APIs of Kafka [47]:

- Producer API: A stream of records can be published to Kafka topics using this API.
- Consumer API: It enables the processing of a stream of records by an application. It also lets an application to subscribe to Kafka topics.
- Streams API: It lets an application to transform the input streams to output streams.

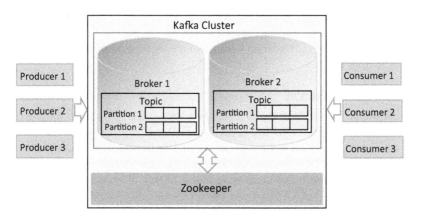

FIGURE 5.10   Architecture of Kafka.

- Connector API: It enables to create and execute reusable producers and consumers that link topics of Kafka to existing applications.
- Admin API: It enables to handle and examine topics, brokers, and various other objects of Kafka.

Kafka is run as a cluster on many servers that span numerous datacenters. Kafka clusters store streams of records in categories called topics. As shown in Figure 5.10, the architecture consists of producers, brokers, consumers, and Zookeeper [9, 13].

- Producers: They send data to the brokers.
- Brokers: Brokers are actually Kafka servers that store the messages. A cluster is composed of multiple brokers.
- Consumers: They consume messages or read data from brokers. A message is a unit of data in Kafka.
- Zookeeper: It is required to apprise producers/consumers about any new broker or failed broker in the Kafka system because brokers can be arbitrarily added and removed from the Kafka cluster. Since Kafka brokers are stateless, therefore, the management and coordination between brokers are achieved via Zookeeper. For each topic, a set of partitions (commit logs), primary and/or secondary, are maintained by each broker. A partition stores a record and each record consists of a key that specifies what partition the message should be delivered to.

The following are the key features of Kafka [13, 47]:

- Decoupling data streams by Kafka enables extremely low latency, making it exceptionally fast.
- Kafka has very high throughput; it supports a huge number of messages per second.

- The partitioned-log model of Kafka enables data to be distributed across numerous servers, thereby achieving high scalability.
- Kafka is highly fault-tolerant because of the replication of partitions over multiple nodes.

5.5.1.6   Apache SAMOA

With traditional data mining tools, it becomes very challenging to do mining of big data streams [48]. Apache SAMOA (Scalable Advanced Massive Online Analysis) is an advanced distributed machine learning and open-source platform for big data stream mining [49, 50] written in the Java programming language. SAMOA basically amalgamates distributed computing and a library of numerous algorithms for big data stream mining. SAMOA can run on various distributed stream processing engines like Apache Storm, Samza, and Flink because of its pluggable architecture. Therefore, it allows developers to reuse and execute their code on several processing engines [51]. SAMOA has implementations of several state-of-the-art classification, clustering, and regression algorithms for distributed machine learning on streams such as Vertical Hoeffding Tree algorithm, CluStream algorithm, and HAMR algorithm. SAMOA also has implementations of various ensemble algorithms like bagging and boosting [30].

5.5.2   COMMERCIAL PLATFORMS

5.5.2.1   Amazon Kinesis

Kinesis is a platform on Amazon Web Services (AWS) for real-time stream data processing at any scale on the cloud [33]. With kinesis, any amount of data can be captured from diverse sources like website clickstreams and application logs, and can be processed and analyzed in real-time for prompt insights into a much simpler and easier way which lets us to respond instantly to critical information about our business and operations. Kinesis uses a check-pointing mechanism for fault-tolerance. With Amazon Kinesis, appropriate tools for a particular application can be chosen with high flexibility and the best part is that the users don't need to worry about managing the infrastructure as Amazon Kinesis is fully managed. Amazon Kinesis can be used for many real-time applications like fraud detection, application monitoring, and generating alerts [31, 52].

5.5.2.2   Azure Stream Analytics

Azure Stream Analytics is real-time analytics and complex event-processing engine that analyzes and processes enormous amounts of high-speed streaming data emerging from numerous sources like websites, social media, and sensors concurrently. Azure Stream Analytics is a server-less job service on Azure and it's not the responsibility of users to manage the underlying infrastructure, servers, and virtual machines; they only need to pay for the processing used for their running jobs. An Azure Stream Analytics job is composed of input, query, and an output. The streaming data coming from a variety of sources can be ingested into Azure from a device using an Azure event hub or IoT hub. Users can analyze the streaming data in real-time by employing an SQL-like query language. For real-time data, visualization users can also build live dashboards [53, 54].

5.5.2.3 IBM Streams

IBM Streams is an advanced real-time computing platform which lets user-developed applications to swiftly consume and perform analysis as soon as the data comes up from numerous real-time sources. Traditional processing entails executing analytic queries based on historic data. IBM's big data platform Streams processes enormous volumes and varieties of data from a multitude of sources while achieving exceptionally low latency, thus enabling decision makers to draw out relevant information for prompt analysis. Therefore, IBM Streams is particularly efficient in domains where traditional batch or transactional systems aren't sufficient, for instance: processing many data streams at high rates is needed, complex processing of the data streams is required, real-time responses to events and changing requirements is needed, and low latency is required when processing the data streams [13, 55].

5.6 COMPARISON OF OPEN-SOURCE REAL-TIME ANALYTICS PLATFORMS

There is no best real-time analytics platform, which is fit for every situation. Every platform is different in its capabilities and features. Therefore, the best possible

TABLE 5.2
Comparison of Apache Storm, Apache Spark Streaming, Apache Samza, and Apache Flink

	Apache Storm	Apache Spark Streaming	Apache Samza	Apache Flink
Processing Model	Stream	Micro-Batching	Stream	Batch and stream
Stream primitive	Tuple	DStream	Message	Data Stream
Latency	subsecs	Few Secs	Subsecs	Subsecs
API Language	Any	Java, Scala, R, Python	JVM languages	Java, Scala, Python
Delivery Guarantees	At-least-once (Exactly once with Trident)	Exactly once	At-least-once	Exactly once
In-Memory Computing	Yes	Yes	Yes	Yes
Implementation language(s)	Clojure	Scala	Scala and Java	Scala and Java
State Management	No	No	Yes	Yes
Maturity	High	High	Medium	Low
Advanced features	No	No	No	Yes
Throughput	Low	High	High	High
Programming model	Compositional	Declarative	Compositional	Declarative

solution depends on the requirements of the user or use case. Table 5.2 shows the comparison among Apache Storm, Apache Spark Streaming, Apache Samza, and Apache Flink. Related works and surveys [2, 3, 13, 14, 15, 28, 32, 35] were consulted for this purpose. We have examined 12 essential characteristics associated with these platforms that can help in choosing a particular platform for certain applications, which are described as follows:

- Processing Model: A real-time processing framework can be a native streaming or micro-batching framework.
- Stream Primitive: It is the key data structure in a stream processing system.
- Latency: It refers to how soon a record can be processed. For tight real-time applications like stock market, latency should be as minimum as possible.
- API Languages: They are the programming languages that can be used to develop applications for this framework.
- Delivery Guarantees: It refers to the accuracy level of the outcomes delivered after a system failure and an effective recuperation of the system contrasted with what the outcomes would be with no system failures. It can be at-most-once: in case of failures, the messages may not be effectively processed, and records may be lost; therefore, it is the least preferred property; at-least-once: the messages are re-processed and re-delivered in case of failure; it ensures that no records are lost. However, for certain operations, it can deliver inaccurate outcomes or exactly once: the messages are processed exactly one time in case of failures. It guarantees that no duplicates are caused and delivers accurate results. Hence, it is the strongest and most desirable property but is quite hard to achieve in all cases.
- In-Memory Computing: It refers to the retaining of data in RAM rather than in disk (secondary memory).
- Implementation Language(s): They are the language(s) that are used to implement the framework.
- State Management: Stream processing frameworks can be stateful; therefore, some mechanisms should be there to preserve and update state information or they can be stateless, i.e. there is no need to store and maintain state.
- Maturity: It refers to the popularity or wide acceptance of the framework.
- Advanced Features: Certain features are needed such as Event Time Processing, Watermarks, and Windowing in case stream processing requirements are complex.
- Throughput: It means the number of records processed within a particular timeframe. For instance, if the requirement is to process huge volume of data and there is no strict constraint on speed or processing time, i.e. no sub-latency is required, in this case, Spark Streaming can be utilized.
- Programming Model: Stream processing systems can be either compositional when developers have to explicitly implement the whole logic or declarative when higher-level abstractions are provided to developers.

5.7 CONCLUSIONS

In this chapter, we defined big data and its associated characteristics. Today, big data is being generated by almost every sector and various solutions have been endowed for the management and processing of this big data. Hadoop-MapReduce is a broadly embraced solution for handling big data. However, Hadoop is natively developed for batch and very high-throughput job execution, i.e. it is appropriate for jobs that process enormous amounts of data over an extended period of time. It doesn't reflect the incessant and boundless nature of data that demands real-time analysis to make the right decisions and take proper actions at the opportune time. These new demands require new solutions. In this chapter, we shed light on various technologies like stream processing and in-memory computing and provided an overview of real-time architectures. Besides, some of the applications were outlined that require real-time analysis. This chapter discussed the state-of-the-art open-source as well as commercial platforms for real-time applications that give organizations the power to build real-time solutions and finally we compared open-source platforms based on various essential characteristics such as latency, throughput, and delivery guarantees that can help in choosing a particular platform for certain applications.

REFERENCES

[1] Praveena A, Bharathi B (2017) A survey paper on big data analytics. 2017 International Conference on Information Communication and Embedded Systems, ICICES 2017.

[2] El Alaoui I, Gahi Y, Messoussi R, et al (2018) Big Data Analytics: A Comparison of Tools and Applications. Lecture Notes in Networks and Systems. Springer, Cham, pp. 587–601.

[3] Oussous A, Benjelloun FZ, Ait Lahcen A, Belfkih S (2018) Big data technologies: a survey. Journal of King Saud University - Computer and Information Sciences 30:431–448. https://doi.org/10.1016/j.jksuci.2017.06.001

[4] Kalra M, Lal N (2016) Data mining of heterogeneous data with research challenges. 2016 Symposium on Colossal Data Analysis and Networking, CDAN 2016.

[5] Pusala M, Salehi M, Katukuri J, et al (2016) Massive data analysis: tasks, tools, applications, and challenges. Big Data Analytics: Methods and Applications. Springer, New Delhi, pp. 1–276.

[6] P. D, Ahmed K (2016) A survey on big data analytics: challenges, open research issues and tools. International Journal of Advanced Computer Science and Applications 7:511–518. https://doi.org/10.14569/ijacsa.2016.070267

[7] Ibtissame K (2017) Real time processing technologies in big data: comparative study. 2017 IEEE International Conference on Power, Control, Signals and Instrumentation Engineering (ICPCSI)., pp. 256–262.

[8] Samosir J, Indrawan-Santiago M, Haghighi PD (2016) An evaluation of data stream processing systems for data driven applications. Procedia Computer Science 80:439–449. https://doi.org/10.1016/j.procs.2016.05.322

[9] Milosevic Z, Chen W, Berry A, Rabhi FA (2016) Real-time analytics. Big Data: Principles and Paradigms pp. 39–61. https://doi.org/10.1016/B978-0-12-805394-2.00002-7

[10]   Mohamed N, Al-Jaroodi J (2014) Real-time big data analytics: applications and challenges. Proceedings of the 2014 International Conference on High Performance Computing and Simulation, HPCS 2014. pp. 305–310.

[11]   Soumaya O, Amine TM, Soufiane et al (2017) Real-time data stream processing - challenges and perspectives. International Journal of Computer Science Issues 14:6–12. https://doi.org/10.20943/01201705.612

[12]   Zheng T, Chen G, Wang X, et al (2019) Real-time intelligent big data processing: technology, platform, and applications. Science China Information Sciences 62:1–12. https://doi.org/10.1007/s11432-018-9834-8

[13]   Dutta K, Jayapal M (2015) Big data analytics for real time systems. Big Data Analytics Seminar. p. 13.

[14]   Yadranjiaghdam B, Pool N, Tabrizi N (2017) A survey on real-time big data analytics: applications and tools. 2016 International Conference on Computational Science and Computational Intelligence, CSCI 2016. pp. 404–409.

[15]   Shahrivari S (2014) Beyond batch processing: towards real-time and streaming big data. Computers 3:117–129. https://doi.org/10.3390/computers3040117

[16]   Kumar T, Srinivasa (2009) Implementation of the Indian national tsunami early warning system. Fostering E-Governance: Selected Compendium of Indian Initiatives. pp. 380–391.

[17]   Balis B, Bartynski T, Bubak M, et al (2013) Development and execution environment for early warning systems for natural disasters. 13th IEEE/ACM International Symposium on Cluster, Cloud, and Grid Computing, CCGrid 2013. pp. 575–582.

[18]   Islam A (2016) Applications of real-time big data analytics. International Journal of Computer Applications 144:1–5. https://doi.org/10.5120/ijca2016910208

[19]   Ta VD, Liu CM, Nkabinde GW (2016) Big data stream computing in healthcare real-time analytics. Proceedings of 2016 IEEE International Conference on Cloud Computing and Big Data Analysis, ICCCBDA 2016. pp. 37–42.

[20]   Sujatha V, Prasanna Devi S, Vinu Kiran S, Manivannan S (2016) Bigdata analytics on Diabetic Retinopathy Study (DRS) on real-time data set identifying survival time and length of stay. Procedia Computer Science. 87:227–232.

[21]   Wang J, Qiu M, Guo B (2017) Enabling real-time information service on telehealth system over cloud-based big data platform. Journal of Systems Architecture 72:69–79. https://doi.org/10.1016/j.sysarc.2016.05.003

[22]   Simmhan Y, Aman S, Kumbhare A, et al (2013) Cloud-based software platform for big data analytics in smart grids. Computer Science and Engineering 15:38–47. https://doi.org/10.1109/MCSE.2013.39

[23]   Simmhan Y, Perera S (2016) Big data analytics platforms for real-time applications in IoT. Big Data Analytics: Methods and Applications. Springer, New Delhi, pp. 115–135.

[24]   Sanla A, Numnonda T (2019) A comparative performance of real-time big data analytic architectures. Proceedings of 2019 IEEE 9th International Conference on Electronics Information and Emergency Communication. pp. 674–678.

[25]   Koffikalipe G, Behera RK (2019) Big data architectures: a detailed and application oriented analysis. International Journal of Innovative Technology and Exploring Engineering 8:2182–2190. https://doi.org/10.35940/ijitee.h7179.078919

[26]   Singh KN, Behera RK, Mantri JK (2019) Big data ecosystem – review on architectural evolution. Emerging Technologies in Data Mining and Information Security pp. 335–345.

[27]   Feick M, Kleer N, Kohn M (2018) Fundamentals of real-time data processing architectures Lambda and Kappa. Ski 2018-Studierendenkonferenz Inform.

[28] Tantalaki N, Souravlas S, Roumeliotis M (2019) A review on big data real-time stream processing and its scheduling techniques. International Journal of Parallel, Emergent Distributed Systems. 35:571–601. https://doi.org/10.1080/17445760.2019.1585848

[29] Tidke B, G. Mehta R, Dhanani J (2018) Real-time bigdata analytics: a stream data mining approach. Recent Findings in Intelligent Computing Techniques. Springer, Singapore, pp. 345–351.

[30] https://en.wikipedia.org/wiki/Apache_Storm

[31] http://storm.apache.org/releases/current/Tutorial.html

[32] Nasiri H, Nasehi S, Goudarzi M (2019) Evaluation of distributed stream processing frameworks for IoT applications in Smart Cities. Journal of Big Data 6: 1–24. https://doi.org/10.1186/s40537-019-0215-2

[33] http://storm.apache.org/Powered-By.html

[34] Arora B (2018) Big data analytics: the underlying technologies used by organizations for value generation. Understanding the Role of Business Analytics: Some Applications. Springer Singapore, pp. 9–30.

[35] Kamburugamuve S, Fox G (2016) Survey of distributed stream processing. Indiana University Bloomington. https://doi.org/10.13140/RG.2.1.3856.2968

[36] Perwej Y, Kerim B, Adrees MS, Sheta OE (2017) An empirical exploration of the yarn in big data. International Journal of Applied Information Systems 12:19–29. https://doi.org/10.5120/ijais2017451730

WEBSITES

[37] https://samza.apache.org/learn/documentation/latest/core-concepts/core-concepts.html

[38] Ficco M, Pietrantuono R, Russo S (2018) Aging-related performance anomalies in the apache storm stream processing system. Future Generation Computer Systems 86:975–994. https://doi.org/10.1016/j.future.2017.08.051

[39] http://samza.apache.org/learn/documentation/0.7.0/introduction/architecture.html

[40] https://samza.apache.org/learn/documentation/latest/deployment/yarn.html

[41] https://github.com/apache/samza

[42] Perwej DY, Omer M, Kerim B (2018) A comprehend the Apache Flink in big data environments. IOSR Journal of Computer Engineering 20: 48–58. https://doi.org/10.9790/0661-2001044858

[43] Carbone P, Katsifodimos A, et al (2015) Apache FlinkTM: stream and batch processing in a single engine. Bulletin of the IEEE Computer Society Technical Committee on Data Engineering 36:28–38.

[44] https://ci.apache.org/projects/flink/flink-docs-stable/

[45] https://kafka.apache.org/intro

[46] www.javatpoint.com/apache-kafka

[47] https://aws.amazon.com/msk/what-is-kafka/

[48] Bifet A (2015) Real-time big data stream analytics. CEUR Workshop Proceedings. pp. 13–14.

[49] De Francisci Morales G, Bifet A (2015) SAMOA: scalable advanced massive online analysis. Journal of Machine Learning Research 16:149–153.

[50] Kourtellis N, De Francisci Morales G, Bifet A (2019) Large-scale learning from data streams with Apache SAMOA. Learning from Data Streams in Evolving Environments. Springer, Cham, pp. 177–207.

[51] Singh MP, Hoque MA, Tarkoma S (2016) A survey of systems for massive stream analytics. arXiv preprint. arXiv:1605.09021.

[52]   https://aws.amazon.com/kinesis/
[53]   https://docs.microsoft.com/en-us/azure/stream-analytics/stream-analytics-introduction
[54]   https://thirdeyedata.io/azure-stream-analytics/
[55]   www.ibm.com/support/knowledgecenter/SSCRJU/SSCRJU_welcome.htm
[56]   https://spark.apache.org/docs/latest/streaming-programming-guide.html

6 Analysis of Government Policies to Control Pandemic and Its Effects on Climate Change to Improve Decision Making

Vaibhav Saini[1] and Kapal Dev[2]
[1]Indian Institute of Technology, Delhi, India
[2]University of Johannesburg, South Africa
Emails: vasa@signy.io and kapal.dev@ieee.org

CONTENTS

DOI: 10.1201/9781003199403-6

6.1 INTRODUCTION

Pandemics and plagues are often responsible for affecting humanity significantly. Millions of people die, economies collapse, and civilizations end. In most cases, the diseases are difficult to contain; hence, it becomes a challenge to governments across the globe. In the last couple of centuries, humanity has seen multiple pandemics like The Great Plague of Marseille (1720–1723) which claimed more than 100,000 deaths, Spanish Flu (1918–1920), which claimed anywhere from 17 million to 50 million deaths [1], and the ongoing COVID-19 pandemic, which has claimed nearly 600,000 deaths and counting.

Another major global challenge that is as catastrophic as a pandemic is climate change, also called global warming, which refers to the rise in average surface temperatures [2] on Earth. It has been scientifically established that one of the main contributors to climate change is the use of fossil fuels [3–6], which expels greenhouse gases into the atmosphere. These greenhouse gases have the ability to trap heat within the atmosphere [7, 8], which has compounding effects such as extreme climate conditions like droughts [9] that render landscapes more susceptible to wildfires [10], excessive rainfall which causes floods [11], and melting ice-caps [12] which increase the sea-levels [13].

Governments and the policies that they make play a key role in dealing with such difficult situations. Governments from all over the world have joined the "The Paris Agreement" in 2016, which is a legally binding international treaty within the United Nations Framework Convention on Climate Change, dealing with greenhouse-gas-emissions mitigation, finance, and adaptation. Similarly, the governments from all over the world are developing policies to contain and control the ongoing pandemic, some of which are with varying levels of success. Researchers from Blavatnik School of Government developed the Oxford COVID-19 Government Response Tracker (OxCGRT) [14] that collects information on which governments have taken which measures, and when. Using the OxCGRT for federal, state, and some city governments of Brazil [15], they assess which parts of the country have placed measures to fulfill the World Health Organization's (WHO) recommendations for relaxing physical distancing measures.

As countries begin to roll back lock-down measures, the questions that come up are how and when do we know it is safe to do so? Our work focuses on creating a comparison between different countries to objectively measure the effect of the policies and providing a multi-national overview of which countries meet four of the WHO's six recommendations for relaxing physical distancing measures using publicly available data. We also measure the effectiveness of these government policies on the environment where they are enacted.

The objectives of the chapter are as follows:

- Providing a multi-national overview of which countries meet four of the WHO's six recommendations for relaxing physical distancing measures using publicly available data.
- Suggesting how countries can meet the WHO's guidelines.
- Studying the effects of the current policies on the environment, by comparing the air pollution levels before and after lock-downs.

The chapter has been structured as follows. In Section 6.2, we will go through methodology, where we discuss the WHO's recommendations for relaxing physical distancing measures, pandemic datasets used in our analysis, their sources, their limitations, and how we use them in our analysis. We then discuss the significance of the Air Quality Index (AQI), air pollution datasets used in our analysis, their sources, their limitations, and how we use them in our analysis to show the changes in the air pollution levels before and after the lock-downs. Finally, in Section 6.3, we discuss our results and provide some general recommendations that can be followed by the governments.

6.2 METHODOLOGY

6.2.1 WORLD HEALTH ORGANIZATION CRITERIA FOR RELAXING PHYSICAL DISTANCING MEASURES

The WHO has outlined six measures governments need to fulfill before rolling back "lock-down" measures. In brief, these are as follows:

1. The number of new COVID-19 cases should be reduced and contained to a level that the country's health system can manage safely. Ideally, transmission should be controlled to the level of clusters of cases. This could be assessed through the continuous decline in the number of cases over a 14-day period, or longer.
2. There should be sufficient public health workers and sufficient health-system capacity to track, test, and isolate all cases, irrespective of the severity of these cases and whether they arise through local transmission or are imported from elsewhere. This requires monitoring the health system, for example, keeping tabs on the number of available ICU beds, so that capacity is not exceeded.
3. In highly vulnerable settings such as hospitals and residential care homes, the main drivers of transmission should be identified and appropriate distancing measures should be put in place to minimize the risk of new cases.
4. Standard prevention measures should be practiced in workplaces, including directives and, where needed, additional capacity to promote distancing of two meters, hand washing, and respiratory etiquette. These measures include teleworking, staggered shifts, and other practices to reduce crowding.
5. Measures should be put in place to minimize and control the risk of import and export of cases. This requires analyzing the likely origin and routes of

imported cases and establishing the means to rapidly detect and manage suspected cases among departing and arriving travelers. Relevant measures include entry screening and the isolation of sick travelers, and quarantining individuals arriving from places with community transmission.

6. All communities should be engaged and should understand the "new normal" that follows a step-wise transition away from strict restrictions, in which behavioral prevention measures are maintained, and everyone has a role to play. This could be assessed through community surveys.

6.2.2 DATA MINING

In order to objectively compare the preparedness of different countries for relaxing physical distancing measures, we have used the open-source datasets as listed in Table 6.1.

6.2.3 DATA ANALYTICS

6.2.3.1 Formulas for Calculating Metrics

Following are some facts and details about the data of Oxford COVID-19 (OxCGRT) dataset. The first eight columns (C1–C8) represent the policy indicators for *closure and containment* policies, such as restrictions in movement access, various demographics, and academic institution closures. The next four columns (E1–E4) represent the *economic* policy indicators such as the provision of foreign aid or income support to underprivileged citizens. The next five columns (H1–H5) represent *healthcare system* policies such as investments into healthcare or the COVID-19 tracking, testing, and treatment infrastructure. Any data collected which cannot be categorized in the above indicators are put as a miscellaneous indicator (M1). This repository contains more in-depth details on the different indicators and their meaning.

6.2.3.1.1 Transmission Under Control

$0.5*(1-(\text{number of active cases/total number of beds available})) + 0.5*(1-(14\text{-day avg. active cases}/14\text{-day avg. global active cases}))$

The first term signifies the capacity of the country to manage the active cases. The second term signifies the relative condition of the country with the world (other countries). For some countries, we do not have the latest number of beds available and hence are excluded from the comparison.

In case we have some missing values in the last 14 days of active cases, we calculate the average for the days for which the data is available. A country with a less number of active cases in the last 14 days and a good ratio of active cases to the number of beds available (less strained healthcare system) will meet the criteria.

6.2.3.1.2 Testing and Tracing

$0.25*(\text{H2}) + 0.25*(\text{H3}) + 0.5*(1-(\text{number of active cases/total number of beds available}))$

TABLE 6.1

The First Column Contains the Names of Data Sources. The Second Column Describes the Type of Data in the Data Sources. The Third Column Describes How the Study Uses the Data Sources. The Fourth Column Describes the Limitations of the Data Sources

Source	Data Type	Usage	Limitations
Oxford COVID-19 Government Response Tracker (OxCGRT)	The Oxford COVID-19 Government Response Tracker (OxCGRT) is a dataset containing 17 indicators regarding policies stated by the governments across 161 countries to counter and contain the pandemic.	Used in evaluating recommendations 1, 2, 5, and 6. See how we use different data points in our formulas below.	• Datasets for some countries are not up to date. • Data for some countries (and their index scores) might be updated retroactively. • Null values are not the same as 0. • Data on financial policies for some countries is not available.
Epidemiological data from the European Centre for Disease Control	Epidemiological data such as the number of COVID-19 deaths and cases, testing statistics, and other country-specific demographic data.	Used in evaluating recommendations 1, 2, and 6. See how we use different data points in our formulas below.	• Datasets for some countries are not up to date.
Worldometer	Worldometer is a provider of global COVID-19 statistics: the number of active, new, recovered, serious, and critical cases, deaths, and tests for more than 100 countries.	Used in evaluating recommendation 1. See how we use different data points in our formulas below.	• Datasets for some countries are not up to date.
Apple Mobility Data	Anonymous mobility data, including the percentage change in the number of people walking, driving, or taking public transit for multiple demographics.	Used in evaluating recommendation 6. See how we use different data points in our formulas below.	• Datasets for some countries are not up to date. • Data from 11 May to 12 May is not available in the dataset.
Google Mobility Data	Anonymous mobility data, including time-based geographical trends across different categories of places such as transit stations, parks, pharmacies, groceries, workplaces, and residential, retail and recreation.	Used in evaluating recommendation 6. See how we use different data points in our formulas below.	• Not all countries are equally up to date.
India Pollution Data	Pollutants data for eight different pollutants for multiple Indian cities from January 1, 2015 to July 1, 2020.	Used to analyze pre-lock-down and post-lock-down pollution levels. See how we use different data points below.	• Not all states are equally up to date. • Some pollutants have more than 50% of missing data points.

Here, H2 represents the most recent value of the testing policy indicator (H2) from the OxCGRT database. Here, H3 represents the most recent value of the contact tracing policy indicator (H3) from the OxCGRT database. The first two terms evaluate the testing and tracing aspects of the metric, respectively. The third term evaluates the health-system capacity to manage the active cases. A country with a good testing policy (open public testing), a good contact tracing policy (complete contact tracing done for all identified cases), and a sufficient number of beds to support the active cases will meet the criteria.

Limitation: This metric does not measure the isolation factor.

6.2.3.1.3 Managing Vulnerable Settings
Due to the unavailability of proper and reliable data, this criterion is not measured currently.

6.2.3.1.4 Putting Preventative Measures into Workplaces
Due to the unavailability of proper and reliable data, this criterion is not measured currently.

6.2.3.1.5 Manage the Risk of Imported Cases
C8/4

C8 represents the most recent value of the international restrictions policy (ban on all regions or total border closure) indicator in the OxCGRT. A country with a good international restrictions policy will meet the criteria.

Limitation: This metric does not measure the risk of exporting cases.

6.2.3.1.6 Communities Are Fully Engaged and Understand
0.5*(H1/2) + 0.5*((100–average mobility percentage)/100)

H1 is the record presence of public info campaigns.

The *average mobility percentage* is the average percent change in the mobility of the masses since March 11, 2020. If the mobility of the masses remains the same before and after the pandemic, then the *average mobility percentage* will be 100 (hence making the value of the second term zero). So, a reduction in average mobility after the pandemic started to result in a higher value.

We use the global transit, retail, work-space mobility reports from Google's mobility database, and driving, walking, or public transit mobility trend reports from Apple mobility reports. Here, the first term evaluates the public awareness campaigns run by the government and the second term evaluates the average reduction in mobility trends after the pandemic started. A country with good public awareness campaigns (nationally coordinated public information campaigns) and a significant average reduction in mobility trends will meet the criteria.

The complete process followed in the study can be visualized in Figure 6.1.

Now as we have discussed the methodology to objectively compare which countries meet four of the WHO's six recommendations for relaxing physical distancing measures, let's discuss the methodology to compare the pre-lock-down and post-lock-down pollution levels across several Indian cities.

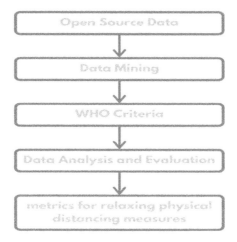

FIGURE 6.1   Analytical process for data preparation.

6.2.3.2   Analyzing Trends in Air Quality Index (AQI) Across Several Indian Cities

Before diving into the methodology, let's first understand the significance of the AQI.

6.2.3.2.1   Air Quality Index

An AQI is a comparison-based index that is used to communicate the current (or forecasted) quality of air. The higher the AQI, the lower is the quality of the air. The AQI consists of six sections: Good, Satisfactory, Moderately polluted, Poor, Very Poor, and Severe. The way AQI is calculated for a city is by measuring the concentration of eight pollutants O_3, NO_2, NH_3, SO_2, PM2.5, PM10, CO, and Pb for a short-period (averaged over 24-hours), and the worst AQI reading represents the actual AQI for the city.

Figure 6.2 lists different AQI categories with the corresponding health impacts consulted from medical experts.

This is a Kaggle tutorial that explains the above AQI calculation process in detail.

Due to increasing cases of COVID-19, the Indian government introduced the first lock-down on March 25, 2020, which aimed to control and contain the number of COVID cases. This meant closing down any non-essential public spaces like places of worship and markets, and only the most essential services that are necessary for the survival of the people like healthcare services, electricity, and water were functional. This had some unintended but remarkable improvements in the air quality and visibility, which led to the scenic view of Dhauladhar Peaks of Himachal Pradesh from neighboring states, which usually are hidden behind a thick film of smog.

AQI	Remark	Color Code	Possible Health Impacts
0-50	Good		Minimal impact
51-100	Satisfactory		Minor breathing discomfort to sensitive people
101-200	Moderate		Breathing discomfort to the people with lungs, asthma and heart diseases
201-300	Poor		Breathing discomfort to most people on prolonged exposure
301-400	Very Poor		Respiratory illness on prolonged exposure
401-500	Severe		Affects healthy people and seriously impacts those with existing diseases

FIGURE 6.2 Different categories of AQI ranges, their corresponding remarks, and possible health impacts with color coding.

Using the vast amount of AQI data available, we will access the objective change in AQI and concentrations of the eight pollutants across different cities all over India.

Our dataset includes data from January 1, 2015 to July 1, 2020, for the following 25 cities of India: Delhi, Ahmedabad, Bengaluru, Chennai, Lucknow, Mumbai, Patna, Gurugram, Jorapokhar, Visakhapatnam, Amritsar, Thiruvananthapuram, Amaravati, Shillong, Brajrajnagar, Talcher, Kolkata, Guwahati, Chandigarh, Bhopal, Ernakulam, Kochi, Hyderabad, Aizawl, and Jaipur.

On analyzing the data, we found that Ahmedabad, Delhi, Bengaluru, Mumbai, Hyderabad, and Chennai have the highest pollution among the 25 cities, as shown in Figure 6.3.

6.3   RESULTS AND DISCUSSIONS

6.3.1   COUNTRIES WITH MAXIMUM READINESS FOR EASING RESPONSE POLICIES

Out of the list of 145 countries that we evaluated, Table 6.2 shows the top 10 countries with maximum readiness for easing response policies. This objective scoring roughly describes how close countries are to achieving four of the six current WHO recommendations. A point to note here is that as the study only focuses on four out of six needed criteria, it is not a perfect scale to judge the readiness if the country is ready to execute a lock-down rollback, but it provides a way to understand where should countries work on more. The OxCGRT data only measures the policies stated by the respective countries, but not how well they are executed. So, it is stressed that countries should follow the specific guidelines stated in the WHO recommendations.

6.3.2   AQI FOR SOME OF THE MAJOR CITIES OF INDIA

Figure 6.4 shows the AQI levels for the following cities from January 2019 to July 2020: Ahmedabad, Bengaluru, Chennai, Delhi, Hyderabad, and Mumbai.
Here are some key facts and observations from the above graph:

- As expected, all the cities, in general, have an extreme level of pollution shown by the high AQI levels.
- The vertical black line shown in the graph marks the time at which the first lock-down was put into place in India.
- We can clearly see a sharp decline in AQI levels across all the cities after the first lock-down was put into place on March 25, 2020.

6.3.3   AQI BEFORE AND AFTER LOCK-DOWN

Now, as we have observed that there is a significant change in the AQI levels, let's see the difference in the AQI levels before and after the first lock-down.

6.3.3.1   Before Lock-down

Before lock-down was implemented, the policies regarding travel restrictions (via cars, buses, trains, and airplanes) and the manufacturing sector (workplace) were not

	City	PM2.5		City	PM10		City	NO2		City	SO2		City	CO
0	Patna	123.500000	0	Delhi	232.810000	0	Ahmedabad	59.030000	0	Ahmedabad	55.250000	0	Ahmedabad	22.190000
1	Delhi	117.200000	1	Gurugram	191.500000	1	Delhi	50.790000	1	Jorapokhar	33.650000	1	Lucknow	2.170000
2	Gurugram	117.100000	2	Talcher	173.920000	2	Kolkata	43.040000	2	Talcher	29.060000	2	Delhi	1.980000
3	Lucknow	111.630000	3	Jorapokhar	149.660000	3	Bhopal	37.490000	3	Patna	22.130000	3	Talcher	1.880000
4	Kolkata	68.690000	4	Bhopal	134.370000	4	Patna	37.490000	4	Kochi	18.400000	4	Bengaluru	1.870000
5	Ahmedabad	67.850000	5	Patna	126.750000	5	Visakhapatnam	37.190000	5	Delhi	15.900000	5	Brajrajnagar	1.860000
6	Jorapokhar	64.230000	6	Jaipur	124.920000	6	Lucknow	33.840000	6	Mumbai	15.360000	6	Patna	1.530000
7	Brajrajnagar	63.880000	7	Kolkata	122.560000	7	Jaipur	33.330000	7	Guwahati	14.660000	7	Ernakulam	1.370000
8	Talcher	63.740000	8	Brajrajnagar	122.550000	8	Hyderabad	28.520000	8	Amaravati	14.260000	8	Gurugram	1.260000
9	Guwahati	63.690000	9	Guwahati	116.600000	9	Bengaluru	28.420000	9	Bhopal	13.440000	9	Kochi	1.170000

FIGURE 6.3 A list showing Indian cities with most concentration of pollutants like PM2.5, PM10, NO_2, SO_2, and CO, respectively.

TABLE 6.2
List of Countries with Highest Indicator Scores Implying Maximum Readiness for Easing Response Policies

Country	Cases controlled	Trace, Test, Isolate	Manage imported cases	Community understanding	Total
Mauritius	0.9997607838	0.9997687327	1	0.7701566952	0.9424215529
Sri Lanka	0.9974754921	0.9159142798	1	0.7374216524	0.9127028561
Morocco	0.9601178137	0.8354329836	1	0.8425264831	0.9095193201
Australia	0.9932771958	0.9100643382	1	0.7159888418	0.904832594
Fiji	0.9992720628	0.9992725509	1	0.6135185185	0.9030157831
Trinidad and Tobago	0.9990442559	0.8323804986	1	0.7305555556	0.8904950775
Jordan	0.9936036636	0.7852959022	1	0.7646866097	0.8858965439
Belize	0.991295215	0.8246292714	1	0.7165242165	0.8831121757
Kazakhstan	0.9326618717	0.9338415313	1	0.62997151	0.8741187282

strict. So, as evident from Table 6.3, the AQI was high, causing serious damage to the overall climate and the biosphere.

6.3.3.2 After Lock-down

After a series of lock-downs were implemented, the policies regarding travel restrictions (via cars, buses, trains, and airplanes) and the manufacturing sector (workplace) were very strict. So, as evident from Table 6.4, the AQI dropped significantly, leading to a better overall climate and a healthier biosphere.

6.4 CONCLUSION

We conclude that every country is unique, and it is not practical to propose a set of one-serves-all policy suites. Each country has a unique geography, natural resources, economy, history, and culture, which deserves its own and unique set of policies that aligns with the country's geography, economy, history, and culture. But there are some basic practices that all governments should follow in the near future to recover from the pandemic in a sustainable way:

- Encouraging businesses to practice work-from-home more frequently. This not only saves time for the employees but also saves the environment from being polluted due to regular travel to the workplace.
- Encouraging virtual education, which eliminates the need for the students to go to a school regularly, saving time and the environment.
- Invest more in science and technology for vaccine discovery and sustainable delivery. As we discover the vaccine, distribution is another challenge ahead of us, and finding ways of distribution that are environment-friendly is something that we should invest in.

AQI Levels

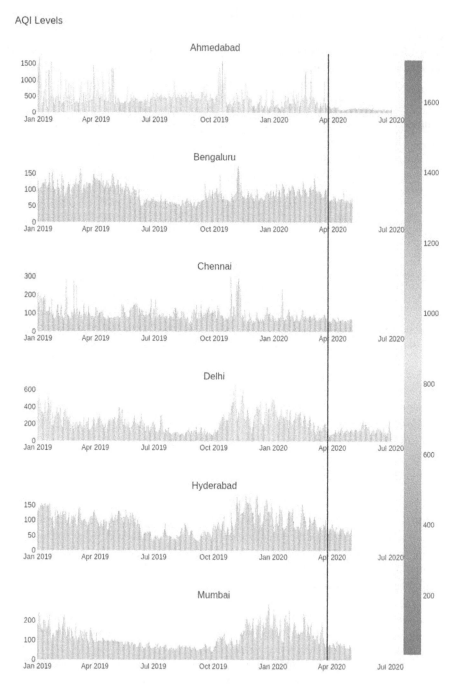

FIGURE 6.4 AQI levels for Ahmedabad, Bengaluru, Chennai, Delhi, Hyderabad, and Mumbai from January 2019 to July 2020, respectively. The vertical black line marks the time at which the first lock-down was put into place in India.

TABLE 6.3
AQI Levels of Different Indian Cities Before
Lock-down Was Implemented

CITY	AQI
Ahmedabad	383.776471
Bengaluru	96.023529
Chennai	80.317647
Delhi	246.305882
Hyderabad	94.435294
Mumbai	148.776471

TABLE 6.4
AQI Levels of Different Indian Cities After
Lock-down Was Implemented

CITY	AQI
Ahmedabad	127.810811
Bengaluru	68.513514
Chennai	62.189189
Delhi	107.270270
Hyderabad	65.675676
Mumbai	73.972973

- Climate change has forced many species to move toward the poles [16]. This is because as the average temperature of the world is increasing, organisms are moving toward colder regions (old normal). This disturbs the biosphere and leads to unwanted consequences at a global level (migration of species). This can be a reason for the introduction of new viruses and bacteria into ecosystems that are not immune to them, causing a pandemic-like situation. This is something that needs to be addressed by governments from all over the world (especially the countries that are major contributors to climate change).

REFERENCES

[1] Pandemic Influenza Risk Management WHO Interim Guidance (PDF). World Health Organization. 2013. p. 19. Archived (PDF) from the original on 21 January 2021. Retrieved 7 December 2020.
[2] Ekwurzel, B., Boneham, J., Dalton, M.W. et al. The rise in global atmospheric CO2, surface temperature, and sea level from emissions traced to major carbon producers. Climatic Change 144, 579–590 (2017). https://doi.org/10.1007/s10584-017-1978-0
[3] Rodhe, H. A comparison of the contribution of various gases to the greenhouse effect, Science 248, 4960, 1217–1219 (1990). DOI: 10.1126/science.248.4960.1217

[4]   Sathre, R. Comparing the heat of combustion of fossil fuels to the heat accumulated by their lifecycle greenhouse gases, Fuel 115, 0016–2361, 674–677 (2014). DOI: https://doi.org/10.1016/j.fuel.2013.07.069

[5]   Montzka, S., Dlugokencky, E. & Butler, J. Non-CO2 greenhouse gases and climate change. Nature 476, 43–50 (2011). https://doi.org/10.1038/nature10322

[6]   Lashof, D., Ahuja, D. Relative contributions of greenhouse gas emissions to global warming. Nature 344, 529–531 (1990). https://doi.org/10.1038/344529a0

[7]   Woodwell, G.M., Mackenzie, F.T., Houghton, R.A. et al. Biotic feedbacks in the warming of the Earth. Climatic Change 40, 495–518 (1998). https://doi.org/10.1023/A:1005345429236

[8]   Kweku, D., Bismark, O., Maxwell, A., Desmond, K., Danso, K., Oti-Mensah, E., Quachie, A., & Adormaa, B. Greenhouse effect: greenhouse gases and their impact on global warming. Journal of Scientific Research and Reports, 17(6), 1–9 (2018). https://doi.org/10.9734/JSRR/2017/39630

[9]   Mukherjee, S., Mishra, A., Trenberth, K.E. Climate change and drought: a perspective on drought indices. Current Climate Change Reports 4, 145–163 (2018). https://doi.org/10.1007/s40641-018-0098-x

[10]   Westerling, A.L., Bryant, B.P. Climate change and wildfire in California. Climatic Change 87, 231–249 (2008). https://doi.org/10.1007/s10584-007-9363-z

[11]   Bronstert, A. Floods and climate change: interactions and impacts. Risk Analysis, 23: 545–557 (2003). https://doi.org/10.1111/1539-6924.00335

[12]   Thompson, L.G. Climate change: the evidence and our options. The Behavior Analyst 33, 153–170 (2010). https://doi.org/10.1007/BF03392211

[13]   Bosello, F., Roson, R., Tol, R.S.J. Economy-wide estimates of the implications of climate change: sea level rise. Environmental and Resource Economics 37, 549–571 (2007). https://doi.org/10.1007/s10640-006-9048-5

[14]   Hale, T, Angrist, N., CameronBlake, E., Hallas, L., Kira, B., Majumdar, S., Petherick, A., Phillips, T., Tatlow, H., Webster, S. Variation in Government Responses to COVID-19 Version 7.0. Blavatnik School of Government Working Paper. May 25, 2020. Available: www.bsg.ox.ac.uk/covidtracker

[15]   Hale, T, Angrist, N., CameronBlake, E., Hallas, L., Kira, B., Majumdar, S., Petherick, A., Phillips, T., Tatlow, H., Webster, S. Oxford COVID-19 Government Response Tracker, Blavatnik School of Government (2020). Available: www.bsg.ox.ac.uk/covidtracker

[16]   Pecl, G. T., et al. Biodiversity redistribution under climate change: impacts on ecosystems and human well-being. Science, 355, 6332 (2003), eaai9214 DOI: 10.1126/science.aai9214

7 Data Analytics and Data Mining Strategy to Improve Quality, Performance and Decision Making

D. Sheema[1] and K. Ramesh[2]
[1]Department of Computer Applications, Hindustan Institute of Technology and Science, Chennai, Tamil Nadu, India
[2]Professor, Department of Computer Science & Engineering, Hindustan Institute of Technology and Science, Chennai, Tamil Nadu, India

CONTENTS

DOI: 10.1201/9781003199403-7

7.1   INTRODUCTION

Data is a collection of facts, such as numbers, words, measurements, observations or just descriptions of things [1]. Data may be in any form like text, image or video. Some of the advantages of data are data redundancy, consistency, data integrity and reduced entry of data, storage and retrieval cost. Using these advantages, a user can manipulate the data in order to perform the necessary tasks. Data can be collected in two ways. They are qualitative and quantitative. Qualitative is descriptive with good articulation, which represents precise information, whereas quantitative is tangible, which represents numerical information and can be measured, i.e. numbers. A qualitative value can be illustrated in two ways: discrete and continuous values. A discrete value can take only the whole numbers and it is countable, whereas continuous data can take only within range and it is measurable. We can collect the data in multiple ways; the best and simple way is direct observation or survey. Survey is defined as the act of examining a process or questioning a selected sample of individuals to obtain data about a service, product or process [2]. Some of the common survey methods are written questionnaires, face-to-face or telephonic interviews, focus groups and electronic (e-mail or website) surveys. In order to collect the data, survey plays a major role. Survey will be effective when the administrator intends to know about customer requirements, opinions, ideas, problem of customer and assesses proposed changes. Hence, the data collection surveys help us to collect necessary information from a certain set of people and lead us to draw an outcome from their opinions, behaviour or knowledge.

The next part of the survey is to analyse the results. There are two predominant ways to analyse: one is extracting with knowledge of statistics and another one is computer software packages. Statistics helps to summarize the results with the help of mean, mode, median, quartiles and standard deviation. The statistically recorded information can be illustrated in different formats like bar graphs, Pie charts, dot plots, line graphs, histograms and pictographs. Using computer software packages, an admin can sort and organize the data to get clear outcomes. In the current scenario, gathering information is an easy task but survey helps to filter and bring out the quality and effective data. Jolliffe writes, "The obvious reason for this is to ensure that the data analysed are correct and complete". The result can be compared and analysed in many ways. The result should be analysed within the survey group or between groups or either of them. Eventually, this would lead us to draw a good outcome.

7.2   DATABASE

Database maintains the current or operational data which assists to record data, and data warehouse upholds the historical and commutative data which support to analyse the data. Database maintains large amount of data in a single software application and it also offers to store, organize and manipulate the information. Using the manipulation, admin can list out the reports like cost estimation, accounting reports and invoices for customers as well. Databases need to be maintained properly in order to provide data consistency, cost-effectiveness, and to avoid problem in disk storage. A good database aim is to reduce redundant data, easily accessible, enable to join table together when it requires and ensure the support, integrity and accuracy of data. The database management system includes the components like hardware, software, data, procedures and database access language. A database designer is responsible for detailed design and data model. A detailed design includes the tasks like tables, indexes, views, constraints, triggers and stored procedures. In order to obtain the quality in data, the performance should be measured precipitously. This makes the designer to improve the data accuracy and it will be supportive to bring out the decisions perfectly.

7.3   DATA QUALITY

Errors are common in all tasks, but it is not advisable in data collection. Data quality is the process of data examining or data profiling. In data profiling, the statistics patterns and informative summaries of the data are maintained. Data quality mainly helps to discern the inconsistencies, finding other abnormalities or anomalies in the data and cleansing of data. Some of the data cleansing activities include removing outliers, missing data interpolation, correcting bad data and filtering out some unwanted data. By using data cleansing, admin can improvise the productivity and recognize mistaken or corrupted data from the database. To promote the data quality in the database, data cleansing plays a major role and by maintaining the previous benchmark, strategies will also facilitate to achieve the objective efficiently. Data cleansing also includes some of the advantages, like improvise the decision making, boost outcomes and revenue, product status, minimize compliance threats, accumulate money and trim down waste. Two major approaches are also there to enhance the quality more effectively; they are quality assurance and quality control.

Quality assurance is process-oriented and focuses on defect prevention, while quality control is product-oriented and focuses on defect identification [3]. Verification and validation in software testing are the best examples for quality assurance and quality control. Quality assurance and quality control use the statistical tools and techniques to find out the product and process efficiencies. KPI (Key Performance Indicator) is a measurable value which helps to identify how effectively a company is achieving the target. In business, a company uses KPI to find out the success rate. Organizations should develop separate and specific KPIs for themselves according to their specifications, so that the data quality can be improved by tracking at each and every stage.

Another method is ROI (Return on Investment); it is a performance which facilitates to find out the efficiency of investment and compares the efficiency with altered investment also. Organizations use parameters to measure the quality check. Quality audit check is also essential which is examined by both external and internal audit teams. Various types of quality audits are in business for checking the excellence; they are internal audits, external audits, second party audits, third party audits, process audits, product audits and system audits. Both company and customer will get benefited by implementing the audit in their business.

7.4   DATA ANALYTICS

Data analytics is a technique for analysing raw data in order to acquire conclusion and enhance the business productivity. There are four types of data analytics:

1. Descriptive
2. Predictive
3. Prescriptive
4. Diagnostic

7.4.1   WHY DATA ANALYTICS?

Data analytics is needed, because it uses analytical and logical reasoning to enhance the business optimization (Figure 7.1).

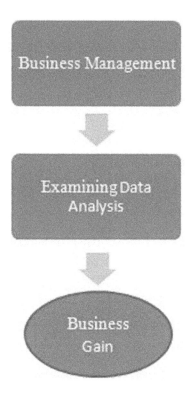

FIGURE 7.1   Process of data analytics.

Some of the important points are as follows:

- Congregate unknown insights.
- Create reports.
- Perform a market analysis.
- Improve business requirements and better decisions.
- Extract knowledge from data to make informed decisions.
- Identify new opportunities in business.

7.4.2   ROLE OF DATA ANALYTICS

Data analytics may include the following roles and responsibilities:

- Data Mining
- Data Cleaning
- Data Analysing
- Create Report with a Visualization Technique
- Database Maintaining and Data Systems

7.4.2.1   Data Mining

Data mining is the process of discovering patterns in large data sets involving methods at the intersection of Machine Learning, statistics and database systems [4].

7.4.2.2   Data Cleaning

It is a foundation process of data preparation which includes analysing, identifying and correcting the cluttered data.

7.4.2.3   Data Analysing

Examine each component of data with the help of analytical and logical reasoning modes. It includes the process of inspecting, cleansing, transforming and modelling data, and supports decision making.

7.4.2.4   Create Report with a Visualization Technique

Report generation translates the analysis into information for future enhancement. Data visualization helps to get eye-catching, high-quality charts and graphs to carry out decisions clearly and concisely.

7.4.2.5   Database Maintaining

Database maintenance is a prescribed set of tasks which helps to improve the database such as performance, free up disk space, check for data errors, check for hardware failure and update internal statistics, fragmentation and integrity check and so on.

7.4.2.6   Data Systems

Data System is a process to collect the symbols in an organized manner and to perform manipulating operations.

7.4.3   SAMPLE DATA ANALYSIS

Data sampling term is a technique of Statistical Analysis used for selection, manipulation and analysis of the concerned data subset to find patterns and trends in a larger data set being examined.

7.4.4   MAIN CHALLENGES USING DATA ANALYTICS

In order to get an effective decision making, the analyst may encounter many challenges. Some of the obstacles are given below,

7.4.4.1   Unstructured Storage of Data

Data sets are not stored in a structured format. It is not easy to search the data and needs a complex algorithm to analyse the data.

7.4.4.2   Data from Multiple Sources

Some of the problems involved in integrating the data from various sources incur solution cost, data replication, data volume, etc.

7.4.4.3 Huge Amount of Data

Organization may collect information at every occurrence, so that it will gather more data in less time. Hence, this process is too time-consuming and surplus data is produced here.

7.4.4.4 Management Pressure

Organizations give inadequate time to complete the work but expect higher returns with a larger number of reports.

7.4.4.5 Data Inadequacy

To make a good decision, the analyst needs accurate input, but the user would prefer to enter the data manually; this can lead to loss of the accuracy of data.

7.4.4.6 Lack of Respective Business Knowledge

Employee does not have enough knowledge and is incapable to analyse the data intellectually. Organizations struggle with powerless hands.

7.4.4.7 Lack of Data Maintenance

The main reasons for improper database maintenance are human error, viruses, malware, hard drive damage, disaster, software corruption, lack of tools and applications, etc.

7.4.5 ANALYTICS FOR BETTER DECISION MAKING

Big data analytics plays a major role to bring better decision making for an organization. The keystones of present strategic business decision making are diverse application and far-reaching use case. Advanced statistical models are also offering valuable insight into data sets and facilitating the business to discover new patterns. Figure 7.2 illustrates the major paces which help to dig the effective ways to get enhanced decision making. The below section facilitates vast opportunities to fetch out the effective business strategies for decision making.

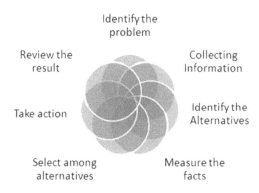

FIGURE 7.2 Effectual paces for decision making.

7.4.5.1   Exploiting Information to Drive Performance

The following techniques are used to drive the quality and performance:

- Gain in-depth understanding and accurate information from the expenditure base.
- Diagnose future prediction.
- Make the economic plan intellectual.
- Benchmark performance.
- Visualization techniques to analyse the data.

7.4.5.2   Utilizing More Concepts of Consumer Patterns

In order to satisfy the customer needs, organizations observe the data from the customer's point of view. Sentiment Analysis can be used to end user necessities. This text analysis technique interprets and classifies emotions of customers in three terms: positive, negative and neutral. This also helps the analyst to find out customer needs and requirements towards brand, product and services. Tools are used for analysis, feedback collection and survey.

7.4.5.3   Governing Risk Through Analytics

Risk management can be categorized into four types: critical, high, medium and low. A Risk Manager will monitor and analyse the business impact when the risk occurs. By implementing the mitigation plan, severity of the problem can be reduced. A Predicative Analysis helps to prevent and provide aids to take precautionary actions, and makes the risk lower. Some of the tools are used for managing the risk; they are BMC Remedy, ServiceNow, Risk Assessment tool, Brain storming, Root cause Analysis, and Risk Register (Figure 7.3).

7.4.5.4   Bottom Line Growth

It is the net revenue of the business after all expenses have been deducted from profits. This analysis highlights net income or net profits.

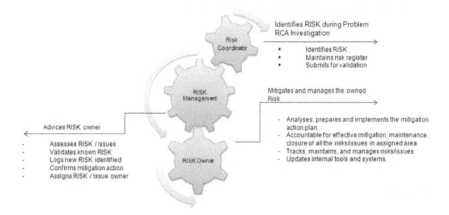

FIGURE 7.3   Examples for managing risk through analytics.

7.4.6 STATISTICAL ANALYSIS

Statistical Analysis is an important analytic technique. This method helps to find out the data based on decisions. The following techniques are essential statistical terms:

- Cluster
- Regression
- Factor
- Neural networks
- Data modelling
- Predicative Analysis
- Cohort

7.4.7 ANALYTICAL TOOLS

The following tools are used in the business for analysing the data to get better accuracy:

- R programming
- Tableau Public
- Python
- SAS
- Apache Spark
- Rapid Miner and so on

7.5 DATA MINING

Data mining is a process which helps to extract the information from a large database. It is also known as Knowledge Discovery in Databases (KDD). Data mining has several steps to extort the required data. Business will collect all the essential data and store into data warehouse. Operational data will be maintained in the database and the historical data will be maintained in the data warehouse. The analyst, managers and organization teams can use warehouse to dig up the needed data (Figure 7.4).

7.5.1 TECHNIQUES INVOLVED IN DATA MINING

- Classification
- Clustering
- Regression
- Association Rules
- Sequential Patterns
- Prediction

7.5.1.1 Classification

Classification is a major technique in data mining which is widely used in various fields and this (Machine Learning) technique is used to predict group membership for data instances [5].

FIGURE 7.4   Stages of data mining.

7.5.1.2 Clustering

Grouping of different data objects is classified as similar objects. One separate group is called a cluster of data.

7.5.1.3 Regression

It is used to forecast a result based on historical data. This technique is used to analyse the relationship between variables. Regression is an important statistical technique for firm applications.

7.5.1.4 Association Rules

An association rule mining is nothing but finding the frequent patterns, correlations, associations or casual structures among the sets of items or objects in transaction database, relational database and other information repositories [6].

7.5.1.5 Sequential Patterns

This technique helps to find statistics-related patterns between the data with discrete values.

7.5.1.6 Prediction

Prediction is used to find and forecast the data facts for future. This technique is used to find future trends, enhance company economy, help to make better decisions, etc.

7.5.2 Functional Areas of Data Mining

Data mining was widely used in many organizations which will help to turn the raw data into valuable information [7]. It aids to develop the market strategy, decrease the cost and also increase the productivity.

Some of the areas of data mining are as follows:

- Application in Retailing
- Application in Banking
- Application in Health care
- Application in Medical
- Application in Insurance
- Application in Education
- Application in Telecommunication
- Application in Market Analysis
- Application in Fraud Deduction
- Application in Manufacturing Engineering

7.5.3 DATA MINING TOOLS

In order to predict the data from a large data set, data mining tools can be used. The following tools are effectively used in the business:

- Rapid Miner
- Oracle Data Mining
- KNIME
- Python
- Orange
- Kaggle
- IBM SPSS Modeler
- Rattle
- Weka

7.5.4 BUSINESS INTELLIGENCE IN DATA MINING

Business intelligence (BI) is the finest method to optimize the decision-making process in business. It has the ability to transform the data into information and information into knowledge [8]. This technique is used to convert the raw data into useful information which turns into profitable business work.

BI involves the data mining, OLAP (Online Analytical Processing) and business reporting to enhance the strategies of business. It can be used for multiple purposes such as measuring performance progress towards business goals, qualitative analysis, data sharing, understand the customer insights and reporting. These tools and techniques provide the solution for the business to upscale the productivity and economy.

- It offers business information quickly and successfully.
- It establishes the metrics to improve the performance.
- It allows the employees to maximize information access for business growth.
- It offers well-configured, quick report generation with data visualization.

- Business helps to screen overall performance of employees, their tasks and output.
- It helps to identify and prioritize the customer, in order to improve the customer satisfaction and enhance the market reputation.
- It protects from online threats such as data breach and malware attacks.

Both data mining and BI are different in definitions but it will provide effective results at the end of the work. BI teams collect filtered data from data mining and this helps to find the root cause and bring out the significance of the data [9].

7.5.5 ANOMALY IN DATA MINING

In data mining, anomaly is used to find the detection of suspicious data in observation or in events, or in rare items [10]. This analysis includes novelties, outliers, exceptions, noise and variations. There are several types of anomalies: Insertion, Update and Deletion.

- Inability to insert data in the database due to lack of other data is called Insertion anomalies.
- The data will become inconsistent due to duplication or redundancy of data or partial update which is called Update anomalies.
- Data cannot be allowed to delete the attribute when the user deletes a record which is known as Deletion anomalies.

In the database, these Insertion, Update and Deletion anomalies are objectionable notions. This drawback can be avoided through a Normalization process.

7.5.6 DATA MINING MODEL

In data mining model, the patterns and trends can be collected and applied to specific scenarios, such as

- Future forecasting
- Risk and probability
- Recommendations
- Sequence Finding
- Grouping

7.5.6.1 Forecasting Method

In order to predict valuable statistics and meaningful data, this method comprises the previously observed value and traditional time series forecast.

7.5.6.2 Risk and Probability

To estimate the risk, probability can be calculated from the targeted customer mailing to diagnose the outcomes.

7.5.6.3 Recommendations
It facilitates to establish which products are likely to be sold together, and create recommendations.

7.5.6.4 Sequence Discovery
Sequence discovery is a data mining technique used to discover the relevant pattern in statistical data [11]. It is also called sequential pattern mining.

7.5.6.5 Grouping
The goal is to arrange a group with similar features and assign them into a cluster.

7.6 DECISION MAKING

Decision-making process is an indispensable part in any organization or business. Management accepts and encourages the rational and sound decisions. In order to get perfect results, the organization practises some methodologies. They are as follows:

- Define the problem.
- Analyse the problem.
- Examine the advantages and disadvantages.
- Finally conclusion to be drawn based on the relevant and closest match of the examination.
- Define corrective and preventive action.
- Monitor the solution.

7.6.1 METHODOLOGIES TO ENHANCE THE DECISION MAKING

By using the above methodologies, a business firm can improve and also lend a hand to progress to another level. In business, 5 whys or five whys techniques are used to find out the cause and effect association underlying a particular crisis.

In Figure 7.5, whys techniques were illustrated with a simple problem. Problem was defined as "Mick reported late to office", so the next question helps to find out, "why he had come late to office"; again, the next question arises, "why did the car break was broke down", so the problem solver will ask the question until discovering of root cause. It can be extended to 6 whys, 7 whys, etc. This is all based upon the problem. But the above problem was concluded only with 5 whys. Also there are other problem solving techniques that can be used such as fishbone and Kepner-Tregoe.

7.6.2 DATA-DRIVEN DECISION MAKING

Data quality is processed as qualitative and quantitative which yields to bring out effective results. There are certain key features which facilitate to understand the quality, such as accuracy, completeness, reliability, relevance, consistency, precision and so on. In order to improve the decision making, the quality traits are used highly (Figure 7.6).

FIGURE 7.5   Five whys.

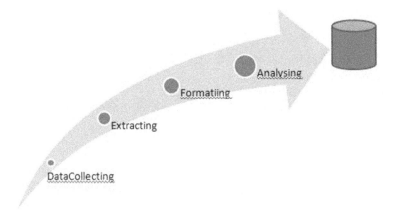

FIGURE 7.6   Data-driven approach.

Data-driven approaches are used in organizations. It helps the business to improve decision-making skills by implementing strategic decisions based on data analysis and interpretation [12],[13]. This method concentrates only to keep the evidence instead of implementing the perceptions, assumptions or instinct one, which can be biased. By using this method, company can examine and organize the data as per the requirement.

7.6.3   RECENT TECHNOLOGIES

In the current world, a user can use many technologies to enhance the data set, extract the quality, get better performance and make decision also. Data can reach the customers easily with the help of techniques such as Visualization, Pattern Recognition, Internet of Things, Machine Learning and Deep Learning techniques [14]. These technologies are embedded with data analytics and data mining to bring out the effective results.

7.7 CONCLUSIONS

Business requires quality data for enhancing the productivity with cost-effectiveness. Data analytics and data mining techniques are highly used in order to bring out the quality data and accuracy. Techniques, methodologies and models are the key elements to improve the performance. These elements enable the user to generate effective decision making in any critical situation.

REFERENCES

[1] Yanqing Duan, John S. Edwards, Yogesh K. Dwivedi, "Artificial intelligence for decision making in the era of big data – evolution, challenges and research agenda", International Journal of Information Management, 48, 2019. pp. 63–71.

[2] Maryam Ghasemaghaei, "Does data analytics use improve firm decision making quality? The role of knowledge sharing and data analytics competency", 120, 2019, pp. 14–24.

[3] Basheer, S., Nagwanshi, K. K., Bhatia, S., Dubey, S., & Sinha, G. R. FESD: An approach for biometric human footprint matching using fuzzy ensemble learning. IEEE Access, 9, 26641–26663, 2021.

[4] Bhatia, S. A Comparative Study of Opinion Summarization Techniques. IEEE Transactions on Computational Social Systems, 8(1), 110–117, 2020.

[5] G. Kesavaraj, S. Sukumaran, "A study on classification techniques in data mining", Fourth International Conference on Computing, Communications and Networking Technologies (ICCCNT), 2013.

[6] Ashwini Rajendra Kulkarni, Dr. Shivaji D. Mundhe, "Data mining technique: an implementation of association rule mining in healthcare", IARJSET, 4(7), 62–65, 2017.

[7] Dahiya K., & Bhatia, S. (2015). Customer churn analysis in telecom industry. In 2015 4th International Conference on Reliability, Infocom Technologies and Optimization (ICRITO) (Trends and Future Directions) (pp. 1–6).

[8] Bhatia, S., Sharma, M., & Bhatia, K. K. (2018). Sentiment analysis and mining of opinions". In Internet of things and big data analytics toward next-generation intelligence (pp. 503–523). Springer, Cham.

[9] Borovska, P., Gancheva, V., "Parallelization and optimization of multiple biological sequence alignment software based on social behavior model", International Journal of Computing, 3, 69–74, 2018. ISSN: 2367-8895.

[10] Bisht B, Gandhi P, (2019) "Review study on software defect prediction models premised upon various data mining approaches", INDIACom-2019 10th INDIACom 6th International Conference on "Computing For Sustainable Global Development" at Bharti Vidyapeeth's Institute of Computer Applications and Management (BVICAM).

[11] Borovska P., Ivanova D., "Intelligent method for adaptive in silico knowledge discovery based on big genomic data analytics", AIP Conference Proceedings, 2048, 060001, 2018.

[12] C.L. Philip Chen, Chun-Yang Zhang, "Data-intensive applications, challenges, techniques and technologies: a survey on big data", 2014, https://doi.org/10.1016.

[13] Ragini, J. R., Anand, P. M. R., & Bhaskar, V. (2018), "Big data analytics for disaster response and recovery through sentiment analysis", International Journal of Information Management, https://doi.org/10.1016.

[14]   Gandhi P., Pruthi J. (2020) Data visualization techniques: traditional data to big data. Data Visualization. Springer: Singapore. pp. 53–74.

[15]   Gandhi K, Gandhi P, (2016) "Cloud computing security issues: an analysis", INDIACom-2016 10th INDIACom 3rd International Conference on "Computing for Sustainable Global Development" at Bharti Vidyapeeth's Institute of Computer Applications and Management (BVICAM), pp. 7670–7673.

WEBSITES

www.mathsisfun.com/data/data.html
https://asq.org/quality-resources/survey
www.diffen.com/difference/Quality_Assurance_vs_Quality_Control
https://en.wikipedia.org/wiki/Data_mining

8 SMART Business Model

An Analytical Approach to Astute Data Mining for Successful Organization

Sharad Goel[1] and Sonal Kapoor[2]
[1]Professor, Indirapuram Institute of Higher Studies (IIHS), Ghaziabad, Uttar Pradesh
[2]Associate Professor, Indirapuram Institute of Higher Studies (IIHS), Ghaziabad, Uttar Pradesh

CONTENTS

8.1 INTRODUCTION

We are staying in tech-age which has made it easy today to collect huge amounts of data. Big retailer outlets are now collecting data of POS, purchasing pattern, share of wallet, etc., each time a transaction happens; credit organizations have a wide range of data on individuals who have or might want to acquire credit, their sources of income and repayment methods; financial organizations have a boundless inventory of data on the historical patterns of stocks, bonds, and different securities, and government offices have information on economic patterns, the climate, social welfare, customer product safety, and essentially all the other things possible. Through this approach the collection of data becomes relatively easy. As an outcome, data is in *PLENTY*. Nonetheless, as numerous associations are presently starting to find, it is a serious challenge to figure out all the information they have gathered.

DOI: 10.1201/9781003199403-8

Besides, technology has given a lot more people the power and to investigate or analyze the data and make decisions based on quantitative analysis. Individuals entering the business world presently cannot pass the entirety of the quantitative analysis to the 'quant jocks', the specialized experts who have generally done the calculations. Quantitative analysis is presently a basic part of their day-to-day stuff. In any case, developing organizations are thinking about this extension approach as a key for upper hand.

8.1.1   BIG DATA APPROACH?

Big data is at the key of the smart revolution for business nowadays. The basic idea behind the phrase 'Big Data' is that everything consumer do is increasingly leaving a digital imprint (or data), which we (and others) can use and analyze to move business ahead and make strategies around Vision & Mission of Organizations. The driving forces of the tech-age world are able to access to ever-increasing volumes of data and our ever-increasing technological capability to mine that data for commercial insights.

8.1.2   WHO IS USING BIG DATA?

Big players like Amazon, Google, Walmart, and Facebook are the companies that are already dealing with more than a million customer transactions each hour and import those into their databases estimated to contain more than 2.5 petabytes of data. These companies are now able to combine this big data from a variety of sources which include customers' past purchases, their mobile phone location, internal stock control records, social media, and information from external sources.

For example a leading telecom company was using big data analytics to predict customer loyalty & satisfaction index and potential churn threat. Based on the airtime usage, VAS services opted, and data usage patterns as well as social media analytics, the company was able to classify customers into different categories. These analytics helped companies to know in advance that a particular set of user is likely to be a churn threat customer. This extremely useful information now helps business/organizations to closely monitor the level of satisfaction of their customers and prioritize actions that help them in preventing them from churn and keep them delighted. These strategies are now widely used by companies for Up-selling and Cross-Selling of Products and Services to gain benefits of incremental ARPUs and strategic business further.

So, big data is changing the nature of business, from manufacturing to healthcare to retail to agriculture, and beyond. The data which can be collected on every conceivable activity means that there are increasing opportunities to fine tune the procedures and operations to squeeze out every last drop of efficiency. Big data is now crucial for the enterprise because of the business value it holds. Data specialists are developing new strategies in analyzing both the legacy data and the tons of new data streaming in.

Both Operations and Development will be responsible for the success of the big data initiative of the enterprise. The walls between the platform, organization, data, and business can't exist at a worldwide scale.

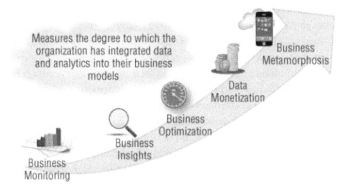

FIGURE 8.1   Big Data Business Model Maturity Index.

8.1.3   BIG DATA MATURITY MODEL

How powerful is our association in collecting, integrating information, and further investigation into our plans of action? The Big Data Business Model Maturity Index is a proportion of how successful an association is at coordinating information and analysis to control its plan of action (Figure 8.1).

The Big Data Business Model Maturity Index provides organizations road maps to integrate data and analytics into their business models by amending current strategies or introducing a new one. The Big Data Business Model Maturity Index consists of five phases:

Phase 1: Business Monitoring – The Business Monitoring phase is the first starting point for almost big data journeys. In this phase, organizations are reviewing and analyzing data inventories and Business Intelligence to monitor the organization's growth to view weekly, monthly, quarterly, or annual reports to know the business successes and problems on a regular basis. Associations have contributed huge time, cash, and exertion to distinguish and archive their key business cycles to make their associations remarkable and fruitful.

Phase 2: Business Insights – In the Business Insights period of the big data, associations need to exploit the financial matters of big data. The financial aspects of Big data empower four new capacities that will help the association cross the analytical gap and move beyond the Business Monitoring stage into the Business Insights stage. It is tied in with utilizing prescient investigation to reveal client, item, and operational bits of knowledge covered in the developing abundance of inward and outside information sources. Organizations forcefully extend their data endeavors by interpreting purchaser experiences and big data into solid activities that drive business development and operational information with inner information. These four major information esteem drivers are access to transactional and operational data, access to Internal and External Unstructured Data, Exploiting Real-Time Analytics, and Integrating Predictive Analytics.

Phase 3: Business Optimization – In the Business Optimization phase, organizations apply predictive (which is likely to happen) analytics to eliminate redundancies, streamlining workflows, improving communication, forecasting the required changes to achieve the business goals and prescriptive analysis (suggests moves that ought to be made) important to improve the focus on key business interactions to convey the significant bits of knowledge or proposals to cutting-edge representatives, business chiefs, and channel accomplices, as well as customers, which further helps organizations to avoid future surprises by taking the corrective action within the time frame specified. This stage additionally tries to impact customer buying behavior and commitment practices by analyzing the client's previous buying patterns, practices, and propensities to convey applicable and actionable suggestions.

Phase 4: Data Monetization – In the Data Monetization stage, companies influence the client, item, and operational experiences to produce new rooms of income openings. They could incorporate packaged data available to be purchased to other companies, integrating analytical data straightforwardly into companies' items and administrations to make, repackaging bits of knowledge to make altogether new items, and administrations that assist associations in entering new business sectors and target new clients or customers. The most noteworthy performing and quickest developing organizations have received information adaptation and made it a significant piece of their methodology. It incorporates more and better experiences for you, your clients and accomplices, smoothes out dynamic and arranging, improves information sharing and coordinated effort among inward and outside partners, strengthens organizations, increases operational efficiency and effectiveness, duplicates and reinforces income streams, provides benefit, improves consistency, and fortifies your upper hand to make new items and administrations, to enter new business sectors and additionally to contact new customers.

Phase 5: Business Metamorphosis – In the Business Metamorphosis stage, the associations look to use the huge information, investigation, and logical bits of knowledge to change the association's plan of action (e.g., measures, individuals, items and administrations, organizations, target markets, the board, advancements, prizes, and motivating forces) in prescriptive examination. It is the stage where associations incorporate the experiences that they find about their clients' utilization designs, item execution practices, and by and large market patterns to change their plan of action. This plan of action transformation permits associations to offer new types of assistance and abilities to their clients in a manner that is simpler for the clients to burn-through and encourages the association to take part in higher-worth and more essential administrations. Furthermore, all the while, these associations will make stage empowering outsider engineers to construct and market arrangements on top of the association's business-as-a-administration plan of action.

Eventually, enormous information possibly matters on the off chance that it assists associations with getting more cash and improve operational adequacy. Models incorporate expanding client procurement, lessening client stir, diminishing operational and support costs, streamlining costs and yield, decreasing dangers and mistakes, improving consistency, improving the client's experience, and so on. In the big data approach, the size of the organization does not hold relevance, as organizations don't need a big data strategy; however, they need good business strategies that can incorporate the apt usage of the big data approach.

8.2 SMART INTEGRATED BUSINESS MODEL (SIBM)

To help the organizations give smart move to their business and also to handle and make the optimum use of big data and data mining techniques, a SMART Integrated Business Model (SIBM Model) is developed. This approach helps organizations to identify where and how to start big data journeys from a business perspective by selecting Unit & Process wise questions. SMART strategies and SMART questions will help in delivering the outcomes as per Budgets and scope of need, to improve performance and harness the primary power of data (see Figure 8.2). This template has already helped many clients to navigate the choppy waters of big (and small) data so that they can reap the rewards without the stress. The blank template is shared for your reference (Figure 8.3).

The aim of the SIBM approach is to help you venture back and ask what are your essential data needs. You can't recognize your data needs on the off chance that you are not satisfactory about your system. Recall that the estimation of information isn't simply the information – it's how you manage the information. So, first of all you need to know what is the specific business requirement; otherwise companies will have the access to all types of data and information but can't develop plan or strategies, which will eventually lead to nowhere, despite investing time and resources to improve business operations.

This you can consider the SMART inquiries to which you need answers. For instance, if your technique is to expand your client base, SMART inquiries that you will require answers to might incorporate, 'Who are presently our clients is the reason,

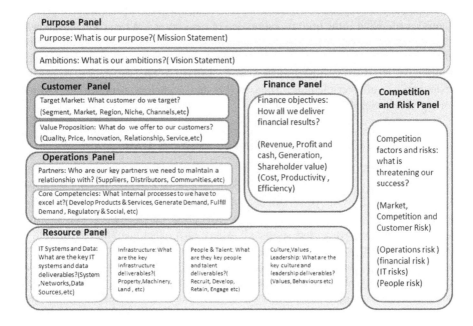

FIGURE 8.2   SMART Business Integrated Model (Blank Format).

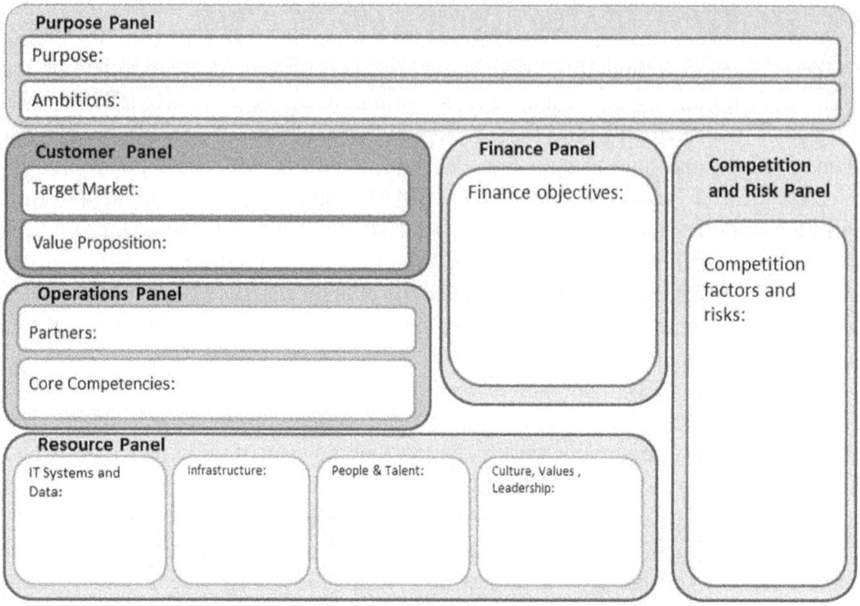

FIGURE 8.3 The SMART Business Model.

it's so critical to begin with system. Assuming you are clear about the thing you are attempting to accomplish?', 'What are the socioeconomics of our most important clients?' and 'What is the lifetime estimation of our clients?' When companies know what to address as per requirements, then data collection becomes easy and faster, which further helps companies to lower Cost & Stress levels to appreciable level. Adopting 'Collect Everything Just In case' to 'Collect & Measure X &Y' to get the desired answer of probable issue. Thus, the big data Technique converts 'Impossible For Us' to 'Absolutely Possible For Us'.

8.2.1 SMART Questions that Help You and Your Team

See the wood from the trees in regards to what's significant and so forth. Comprehend the importance of the information looked for in light of the fact that SMART inquiries show to everybody what your organization's greatest concerns are? Open correspondence and guide conversation. Settle on better proof-based choices. The SMART Business Integrated Model has six panels where each panel has SMART questions that will guide and provide a road map to implement strategic plan of action in any organization. Use your SMART questions to guide your data needs so as to deliver relevant and meaningful information rather than being overwhelmed by the data. Each panel provides an analytical approach that triggers four to five SMART questions per panel. These questions will then form the basis of big data analytics strategy. These panels are as follows:

1. The Purpose Panel.
2. The Customer Panel.
3. The Finance Panel.
4. The Operations Panel.
5. The Resource Panel.
6. The Competition and Risk Panel.

8.2.1.1 The Purpose Panel

The purpose panel gives an inspiring system or in general setting in regards to the corporate methodology or what your business is focusing on or looking to accomplish. Organizations don't typically create SMART inquiries for this board; yet its job is more to set in general setting and bearing. This can best be accomplished by itemizing your central goal and vision proclamation – each doing an undoubtedly extraordinary work.

Your statement of purpose should be clear, the compact mission statement setting out why your association exists. A statement of purpose should impart your aims capably, giving a guide to control the activity plan and dynamic as you endeavor toward the essential objective. It is fundamentally an inward record intended to spur partners and characterize the critical proportions of authoritative achievement which incorporates your intended interest group, what items or administrations probably you segment for subscribers along with the USP of Products or Services.

Your vision statement defines a vivid image of what you want your business to be at some point in the future; based on your goal and aspiration, it will give your business a clear focus and a direction to internal and external stakeholders. It inspires the internal employees to give their best by a strong and meaningful vision statement. The Vision Statement of Business should focus on Corporate Values and behaviors that adhere and display by Stakeholders.

8.2.1.2 The Customer Panel

The customer panel helps retailers and merchants to obtain a reasonable image of purchaser conduct and their moving patterns, including shopping families, their buy practices, what their identity is, the place where they shop, and what they purchase? It prompts you to consider the amount you presently think about the clients? Is your methodology focusing on the clients? What you may have to discover to convey according to your essential targets? The customer panel is classified into two sections: target market and incentive.

The initial segment of the customer panel deals with the methodology (counting your main goal and vision) like what is your objective market? Is it true that you are intending to focus to a specific fragment – if so why and what do you think about that portion? Is it true that you are focusing on the client of a specific geographic district or explicit segment? Provided that this is true, what do you need to think about those likely clients to improve the odds of progress? What are the different modes that you will use to focus on your expected clients?

The second part of the customer panel hints you to clarify your value proposition or what you are going to offer your target customers. Why are these potential customers going to select you to buy products/services from you leaving aside

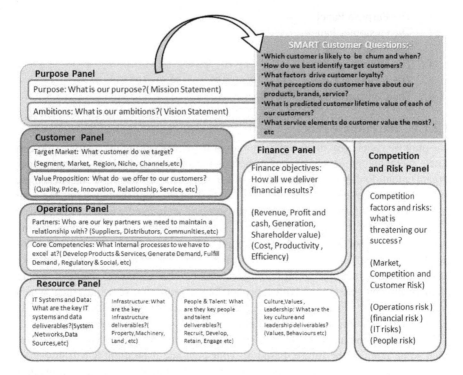

FIGURE 8.4 SMART Customer Questions: Deriving Key Performance Questions from SMART Integrated Business Model.

competition? Do you think they will value for your quality, price, innovation, service, or something else? What will contribute to customer satisfaction and loyalty? Do you know? These SMART customer questions will help you to implement the strategy thinking about your customers (Figure 8.4).

8.2.1.3 The Finance Panel

The finance panel helps the retailers and merchants to get a reasonable image of purchaser conduct and their moving patterns, including shopping families, their buy practices, what their identity is, the place where they shop, and what they purchase? It prompts you to consider the amount you presently think about the clients? Is your methodology focusing on the clients? what you may have to discover to convey according to your essential targets? The customer panel is classified into two sections: target market and incentive.

Initial segment of the Finance Panel, thinking about the methodology (counting your main goal and vision) what is your objective market? Is it true that you are intending to focus to a specific fragment – if so why and what do you think about that portion? Is it true that you are focusing on client of a specific geographic district or explicit segment? Provided that this is true, what do you need to think about those likely clients to improve the odds of progress? What are the different modes that you will use to focus on your expected clients?

8.2.1.4   The Operations Panel

The operations panel prompts you to consider what you really need to do inside to convey your system and what you may have to discover. There are two segments of the operations panel – partners and core skills.

Initially, the association needs to think about which providers, distributors, accomplices, or different middle people are essential in conveying your technique. Do you at present work with these individuals or will you need to make the connections? On the off chance that the connections effectively set up, how solid would they say they are at the present time?

Likewise you need to consider what center capabilities you need to fuse, on the off chance that you will execute your arranged methodology. Are there any holes? Assuming this is the case, how simple is it going to be to fill those gaps?

Do you know, or would you say you are making suspicions? What cycles should be embraced and improve on the off chance that you are to convey what is the need of your objective market? When you are sure about the center and individual components featured in the client, money and activities panels, you need to check how they sway moderately on one another. We ought not to fail to remember that the client, account, and tasks are the center of the business and they should be taken together. Contemplating your activities comparable to your methodology and how it dovetails with your clients and account will trigger SMART tasks questions – or questions you need answers to (see Figure 8.5). These inquiries will at that point

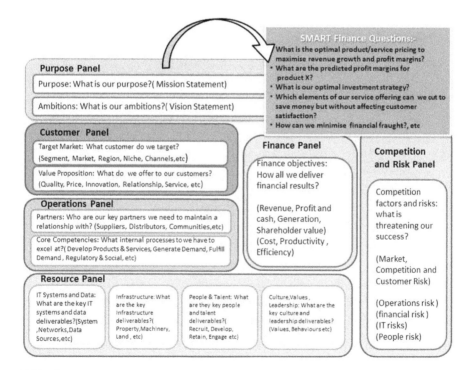

FIGURE 8.5   SMART Finance Questions: Deriving SMART Questions from your SMART Model.

shed light on what sort of information you should gather to respond to those inquiries.

8.2.1.5 The Resource Panel

The resource panel helps you in choice of adept assets, which helps in conveying strategies and the path forward for associations. There are four segments of the resource panel: IT Systems and information, foundation, individuals and ability and societies, values & leadership.

Taking one by one you need to consider: What IT frameworks and information sources would you say you will have to convey your technique? What framework property, hardware, or plant – would you say you will require? What are your Human Resources and Talent pool necessities? Do you have the perfect individuals and if not, would you be able to discover them? Do you need to prepare your current labor or require new ability/group? Lastly, what are the key culture and initiative expectations that will empower this procedure? Pondering the different assets, you will require admittance to comparable to your system will trigger SMART assets questions – or questions you need answers to (see Figure 8.6).

Again, these questions will then shed light on what type of data you will need to collect in order to answer those questions.

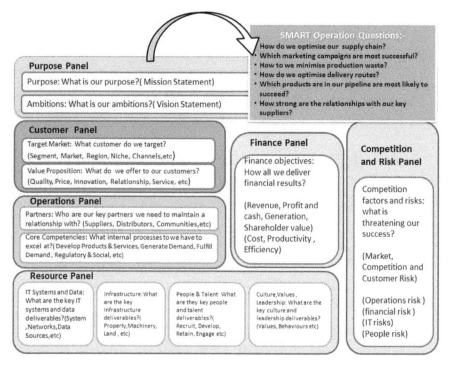

FIGURE 8.6 SMART Operations Questions: Deriving SMART Questions from your SMART Model.

8.2.1.6   The Competition and Risk Panel

The competition and risk panel prompts you to consider what rivalry you will be facing as you look to convey your system and what chances you may look en route. This opposition and hazard viewpoint is the point of view that is frequently absent from methodology maps but it represents a genuine expected danger to effective vital execution. Taking into account what you are looking to accomplish, who is your principle rivalry and why? What is possibly compromising your prosperity? Are there a particular or undiscovered market, client, rivalry methodologies, or administrative standards that may wreck your system? What are the other operational, monetary, or ability-related dangers you face?

Considering your opposition and the different dangers you could face will trigger SMART rivalry and hazard questions or questions you need answers to (see Figure 8.7). These inquiries will at that point shed light on what kind of information you should gather to respond to those inquiries (Figure 8.8).

The truth is that most organizations won't ever have the cash, innovative capacity, or the ability to interminably mine immense, chaotic data sets in the desire for uncovering a newsworthy chunk. Furthermore, that is OK. Zeroing in on SMART inquiries permits us to disregard big data and puts the spotlight on SMART data so that we work out precisely what we need.

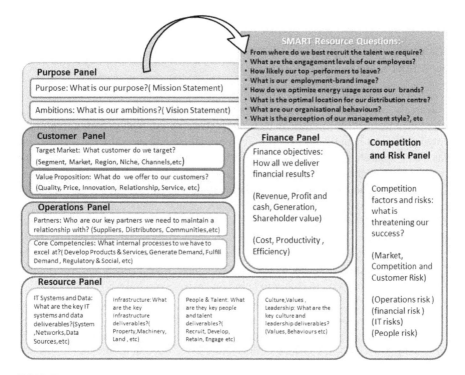

FIGURE 8.7   SMART Resource Questions: Deriving SMART Questions from your SMART Model.

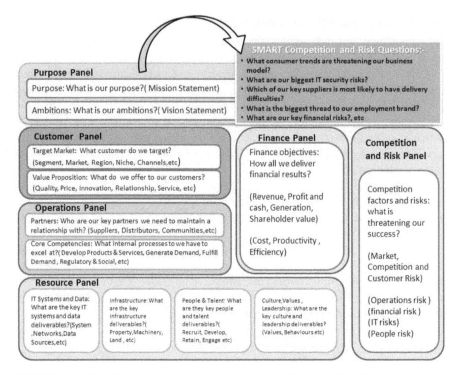

FIGURE 8.8   SMART Competitive and Risk Questions: Deriving SMART Questions from your SMART Model.

Scarcely organizations have the opportunity, tendency, or assets to gather the unlimited information to address they didn't have to get some information about replying. It's not beneficial or useful. Truly we are so entranced by the information that we've neglected that really the inquiry is substantially more significant than the appropriate response the information may, could, or ought to give. Also, you need to understand what addresses you need answers to before you plunge into the information – large or something else.

Brilliant inquiries permit you to express precisely what you need to know with regard to every one of your essential goals so you can focus on what will be deliberately significant and dispose of the rest. These inquiries thus assist you with distinguishing your data needs so you can recognize what metrics and information (M in SMART) you need to gather to help you answer your SMART inquiries.

8.3   CONCLUSION

For driving associations, enormous information gives once in a blue moon business freedom to construct key capacities, abilities, and applications that improve key business measures, drive a seriously convincing client experience, uncover new adaptation openings, and drive serious separation. Keep in mind: purchase for equality, yet work for serious separation. At last, associations have been instructed to think less expensive, more modest, and quicker; yet they have not been educated to think in an unexpected way, and that is actually what's required on the off chance that you need to abuse the enormous information opportunity. Comprehend what has changed and figure out how to think distinctively about how your association uses information and investigation to convey convincing business esteem.

As evident from the above study, the use of big data approach in local and global business is imperative, as it is a single and holistic approach which encompasses all relevant domains required to run business effectively and efficiently. The big data approach is the overall management of micro and macro factors, which can be utilized and further worked upon to identify market space, consumers and their behavior, competitive strategies, workforce management, and overall long-term vision and mission of any business by focusing on relevant data collection and making strategies.

REFERENCES

[1] www.pdfdrive.com/big-data-for-beginners-e200378200.html
[2] www.pdfdrive.com/big-data-using-smart-big-data-analytics-and-metrics-to-make-better-decisions-and-improve-performance-e181764605.html
[3] Big Data Analytics (PDFDrive.com).
[4] Basheer, S., Nagwanshi, K. K., Bhatia, S., Dubey, S., & Sinha, G. R. (2021).FESD: An approach for biometric human footprint matching using fuzzy ensemble learning. IEEE Access, 9, 26641–26663.
[5] Bhatia, S. (2020).A Comparative Study of Opinion Summarization Techniques. IEEE Transactions on Computational Social Systems, 8(1), 110-117,
[6] Bisht B, Gandhi P,(2019) "Review Study on Software Defect Prediction Models premised upon Various Data Mining Approaches", INDIACom-2019 10th INDIACom 6th International Conference on "Computing For Sustainable Global Development" at Bharti Vidyapeeth's Institute of Computer Applications and Management (BVICAM).
[7] Gandhi P., Pruthi J. (2020) Data Visualization Techniques: Traditional Data to Big Data. In: Data Visualization. Springer, Singapore. pp 53–74.
[8] Sheikh, R. A., Bhatia, S., Metre, S. G., & Faqihi, A. Y. A. (2021). Strategic value realization framework from learning analytics: a practical approach. Journal of Applied Research in Higher Education.

9 AI and Healthcare

Praiseworthy Aspects and Shortcomings

Ashay Singh¹ and Ankur Singh Bist²
¹Machine Learning Engineer, US Tech Solutions
Pvt Ltd, India
²Chief AI Data Scientist, Signy Advanced Technologies, India

CONTENTS

9.1 INTRODUCTION

Deep Learning has grown at a breakneck pace over the last decade, and now it has become omnipresence. Some people tend to argue on this argument, but in today's world, more and more domains are seeing its dominance or are accepting it with open arms. Slowly it has resulted in this thing many would regard as their "Technical God", as we often turn to it for many of our technical problems. AI technologies have been widely used in machine vision, self-driving cars, robotics, IoT, Healthcare, etc. Different aspects of AI have been widely used in the healthcare domain. To exemplify its power, in a test held in the Netherlands between AI and 11 expert pathologists to predict Breast Cancer, the AI algorithm was able to solitarily beat the team in the test of the disease on various patients by a significant margin. Advancements in computational resources and AI pipelines escalated the possibilities to automate the healthcare process, but still AI is not mature enough to replace doctors. The current role of AI is to assist doctors in making better decisions. Few systems with AI and robotics capabilities assisted doctors beyond decisions [1]. With the help of big data analytics, useful information can be extracted to support various clinical operations. Various startups are arising and evolving in AI and Healthcare domain as it has a bright future and is the need of the hour. Lots of research articles have been written in the past toward this direction at a fast pace. As per the recently published article [2], there is significant growth in research articles in this domain [2] (Figure 9.1).

DOI: 10.1201/9781003199403-9

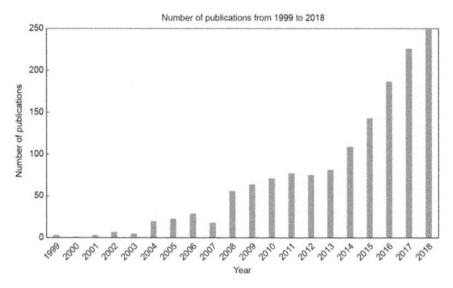

FIGURE 9.1   AI and biomedicine research publication growth.

Just as in real life, sometimes our prayers are heard soon, i.e., we find a suitable algorithm for the problem, and we reach a solution, and sometimes it takes longer to reach our goals. The end goal of AI in Healthcare is to shift the current thread of this domain to the next level, where Healthcare will be more accurate, preventive, and personalized. In this chapter, we have highlighted four crucial aspects of AI and Healthcare.

1   Medical Imaging and Diagnosis.
2   NLP to the Rescue.
3   Deep Learning in Drug Discovery.
4   Health Monitoring Device.

9.2   HOW AI WORKS IN HEALTHCARE?

The functioning of AI is not very different from the pipeline used in other fields. The complete flow of the ML production system varies slightly or substantially from process to process [3]. The difference is dependent on the application and the subdomain it is being applied. The following are the necessary steps followed:

1.   The specification for the real-world problem is decided. Then, the feasibility check is also done at the start to check for the possible outcomes that would result from the pipeline.
2.   After the collection of the relevant data for the project, anonymization is done. It is to ensure the privacy of all the patients who participated in the study. If the recorded data is unlabeled, which is often the case, then annotation is done to make it useful.

3. Now the selection of the best model is made, which is subject to the task at hand. It can be a CNN model for classification, LSTMs for textual data, or other suitable architectures as per the need.
4. The model is then deployed or served for inference and thus moves through the production stage.
5. Now it is verified by various Domain-experts on factors such as task efficiency, model latency, usefulness, and many more.
6. This step is often different from other fields. After all the previous steps, the system is sent for regulatory approvals and permission for clinical trials.
7. After permission is granted, the model is studied in a real-world scenario, i.e., testing in the real-life case for the purpose it was built.
8. After the model is marked as up to the mark, it is made acceptable for the medical staff. It is done by giving them proper training in using the system as it would be they who would be using it in the long run.
9. The model is kept under check to ensure the desired level of performance and check for any possible updates required over time.

All the healthcare AI models follow the above steps, although they all vary to some extent in the context of application for which they are to be used.

Figure 9.2 shows the most common components of any AI-based system that is being deployed in the healthcare domain. The system takes all kinds of inputs, whether they be voice, image, video, graphs, and uses them to make necessary decisions. It thus makes use of the algorithms that are tailored for any specific tasks and provides various useful insights. For example, speech data can be directly used by converting it into various spectrograms or also be converted into text. The raw speech can be used to map various features such as person pitch, mean amplitude, and many sophisticated features [4]. These features can ensure a person's mood or health. AI can help identify patterns in the voice of a sick person to an average person as the regular speaking pattern changes to some degree. This feature is also used by doctors but not on such a large scale, where the number of diagnostics per minute has crossed the human limit by a significant margin. The text can be used for a general chat with any bot, which can suggest some cures based on the knowledge it acquired over time. In general, AI systems can be made to run autonomously or with some human interventions.

Nowadays, robotic surgery has also become very popular [5]. In this kind of surgery, robots are used as assistants or through various control systems. AI has seen all this boom in the field due to its ability to pace up the process. These systems are not necessarily there to replace the doctors but to assist them or take away some of the repetitive and time-consuming tasks to give them more time for critical tasks.

Even away from all the surgery or medication-related tasks, AI also helps the healthcare industry by handling the vast bulk of data they produce regularly.

9.3 MEDICAL IMAGING AND DIAGNOSIS

Medical Imaging technique is solely dependent on feature extraction skills. If done correctly, these techniques can have significant clinical potential for the healthcare

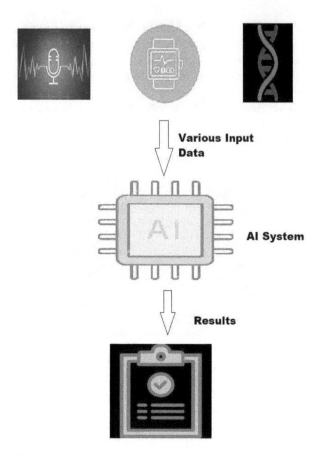

FIGURE 9.2   General flow of AI-based healthcare system.

industry. The most crucial milestone in image classification tasks was the 2012 ImageNet Large-Scale Visual Recognition Challenge (ILSVRC) [6]. In this, the aim was to identify a large corpus of images belonging to various classes. It was after the results of this challenge that people got to know the prowess of Deep Learning. With many such events, Deep Learning became the de facto technique for any image-related task. This change was all thanks to the fantastic image visualization and processing power of Convolution Neural Networks [7].

Figure 9.3 portrays a simple pipeline used for any image-related task. It starts with input, which is usually an image, video frame, or signal spectrograms. Then a Deep Learning-based model, preferably a CNN-based model, is used for feature extraction. It can effectively extract various features from the images and that too using fewer parameters. CNN can extract in-depth features without the need for many parameters, so they are handy for image-related tasks. Then these features are flattened and sent as an array of values to the neural network classifier. The classifier learns these features and tries to make predictions based on that. It looks for similar features when a new image is presented to it and classifies it accordingly. Then a report is generated based

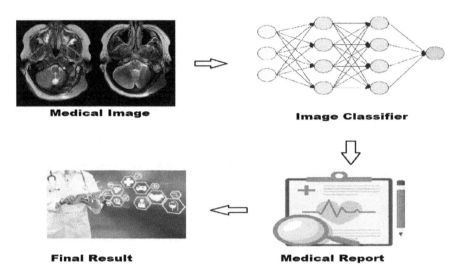

FIGURE 9.3 Working of a basic image diagnostic pipeline.

on outputs received. A doctor sees this report and proceeds for any further tests or treatments if necessary. So, a DL pipeline is beneficial to speed up the healthcare process.

Deep Learning is used for various tasks in image-based medical applications. The simplest is the use of a DL-based image classifier that classifies a patient as sick or well. It can also be used for segmentation tasks, which include the identification of regions where some abnormalities occur.

The model divides the image into various segments and marks the ones that show any abnormality. One of the most used architecture for this task is U-net [8].

Segmentation is used for CT scans and tumors detection. It also helps to obtain a better result by enabling the enhancement of medical records. MRI scan is a very time taking process, and any movements or mechanical and human error can result in a noisy image. With the help of image enhancement techniques, these kinds of noise can be filtered, and a better image can be obtained. Some of the techniques used for this are denoising autoencoders and Generalized Adversarial Networks (GANs) [9]. GANs are also used to clean various MRI-based artifacts. One of the other applications of these networks is a super-resolution technique which can improve the quality of an image by a significant margin. This technique can help work on a poor-quality image as it can now be converted into a better resolution image that is better for use.

Although with all these advancements and levels of performance, AI still is far from being a fully autonomous system in healthcare. Even today, it is hard to believe in it entirely as the features used might not be the ones to look for in a patient. Nevertheless, even with DL being a black box technique, no one can deny the wonders it can do in the field.

9.4   NLP: A NATURAL SOLUTION TO HEALTHCARE PROBLEMS

Natural Language Processing (NLP) is the ability of a machine to process human speech and extract some meaning out of them. The input can also be textual. NLP is quite often used on a large corpus of unstructured data to make it structured or semi-structured. It also helps in the improvement of the clinical documentation process [10]. It saves time by allowing doctors to focus more on practical care of patients than taking and arranging all the medical documents.

Figure 9.4 shows the use of an app to keep in touch with a patient. Chabot can also be designed to remind the patient of his prescriptions and to book appointments. In it, the Chabot is talking to the user. It is done to check for patient's health and needs. It provides correct replies and responds accordingly. Also, when it is unable to provide proper replies to the user, it presents two options that can help the user to proceed further with his needs.

It works on the principle of encoder-decoder based models. In this, usually, a Swq2Seq based model is used. The encoder receives the input stream of the text as a vector representation of the original text [11]. Then, the encoder produces an intermediate representation of this text. This representation is used by the decoder to convert them into words one at a time. The RNN-based models are not suitable for very long sentences. For this purpose, an LSTM network is mostly used.

Speech-based systems are also used for various purposes [12]. They can be transformed into spectrographs to record various features of any person's voice. It can be used for health diagnostics of a patient, and his behavioral trails can be used to ensure his well-being. Also, the speech to text engines is used to convert the voice of a person to text, and further processing is required as necessary.

In this file, the performance of AI boils down to the ability of a computer to interpret the human language. It is one of the limitations that NLP models cannot remember

FIGURE 9.4   Application of Chabot in healthcare.

too old history. Also, the understanding of human language has not made tremendous improvement. Still, DL-based systems struggle to extract meaning out of the sentence that is to a preferable degree. NLP can make some decisions about the intent of the user, mood, and others. In the case of long sentences, the complete meaning is hard to extract without leaving some information out of the input. However, with the latest advent of frameworks like OpenAI's GPT-2 [13], and GPT-3 [14], NLP has evolved and has come much closer to human-level understanding. GPT-3 can generate the human-level composition of texts, i.e., texts written at the level of humans. It has about 175 billion parameters in the network. These parameters enable the network to learn effectively.

GPT-3 is said to be able to complete the incomplete sentences and many other tasks. However, as mentioned in [15], these state-of-the-art models like GPT-2 and BERT still cannot perform all the tasks that humans can with some acceptable level of performance. In the paper, the authors tested the model on ordinary level inference. They tested the model's ability to answer passage-based answers. Nevertheless, even on simple passages, the models struggled to make sense of a follow-up sentence. It means it was unable to connect meaning between two sentences that were related somehow correctly, and the task was straightforward for a human to do. In their study, they found that the models could only reach half the level of human performance (human>95% and models<48%). This study suggests that NLP models are still lacking in many of the tasks but seeing their evolution over time, they might be able to close this gap soon.

9.5   DEEP LEARNING IN DRUG DISCOVERY

Drug discovery is the process of proposing a new possible medication contender for any possible disease. It is one of the biggest industries that work for the social benefit of all. It is a billion-dollar industry but with little promise for high returns. It is because earlier methods were slow to produce the results and were not sure shot to result in any beneficial outcome. Penicillin [16], a commonly known drug used against bacteria, was also discovered in an accident. When its inventor Sir Alexander Fleming, a Scottish researcher, was leaving for his vacations, he accidentally left the petri dish in which he was growing the bacteria uncovered. When he returned from vacation, he found that some of the bacteria were destroyed by a fungus growing around it. He researched more on this fungus and later found that it was quite effective against the bacteria. So, in this surprising way, Penicillium notatum was discovered, which is still regarded as an excellent medicine for fighting against bacteria [17]. Drug discovery was a very time-consuming task and often did not produce satisfactory results. It was also one of the main reasons why more and more advancement was made for some diseases and not all.

All the above steps confirm that the process of designing a new drug is a very tedious task. It is also one of the main reasons why drug or treatment for only 10–20% of diseases are improving a lot over time. The other being the fact that companies are not getting required funds for research of other diseases. This action has resulted in an imbalance between available treatments for different diseases. For some, like Cancer, the treatments available are for several stages of disease progression, but for others,

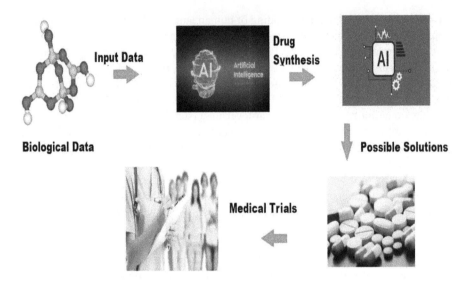

FIGURE 9.5   Traditional flow of the drug discovery process.

only some essential treatments are available. It is a direct result of a decrease in the company's revenue from investing in other treatments that are more time taking. It is where AI comes into the picture. AI has helped to speed up the process of drug discovery with its powerful computational capabilities and efficiency in doing the same (Figure 9.5).

In a typical AI-based drug invention pipeline, the first step is to take account of all the disease-related compounds. Then, the properties of various drugs are studied, and some possible solutions are discovered that are weakly related to the cure of drugs. Then, features from all these drugs are extracted, and new and better drug structures are proposed, which are mostly better at tackling the problem than previous ones. Once a satisfactory drug is received, it is tested for any possible side effects and toxicity. When the testing is completed at a preliminary stage, and the drug is marked safe for human use, it is moved to the medical trial stage. The drug discovery process [18] includes an evaluation of how a drug interacts with any specific target protein. Since all living beings are primarily made up of various proteins, so changing how an external protein interacts with them changes the structure. This change in structure can be both helpful and harmful based on the new properties as per the combination of protein results. For this protein, structures are encoded as strings and classified during testing based on specific properties.

For a drug to be cure of disease A it must have property "T", so classification can be used to check if the property "T" is present in protein k or not. It can also be treated as a regression problem to check similarity to any feature or rank essential features.

AI has brought about massive changes in the field. One of the best examples could be of currently ongoing COVID-19 Pandemic [19]. Earlier, the drug discovery process took about a decade to become promising for use. With the power of Deep Learning, many institutes are proposing a fully-fledged and possible solution to be active within a year. That is a ten times boost, and the performance still grows. In this, many are using a different part of the drug discovery process that is called the *De Novo Drug Design* [20] process. In this, the main aim is to design drugs that have some specific chemical or physical properties. It could be the ability to bind to a particular protein strongly and have excellent water solubility. Achieving this was also regarded as a difficult task as we want to obtain drugs based on our own desired properties. It might also happen that no other drug exists and cannot be made with such conflicting properties. However, with the advancement in Generalized Adversarial Networks and Variational Encoder models, this field has also seen a significant boom over time. Many drugs proposed as a possible cure for the novel coronavirus were built using networks making use of variational autoencoders in combination with other techniques. These drugs were said to attach to the outer boundary of COVID and stop its interaction with human cells, thereby preventing its spread. Even with all these changes that AI has brought in the field, the time is still far when we can trust the drugs invented using it with very minimal testing. It keeps amazing the researchers with its power, and that time also might not be too far in the future.

9.6   HEALTH MONITORING USING AI

Health Monitoring evolved with the evolution of various smart gadgets over time. Earlier for any kind of physical checkups, it was necessary to visit nearby hospitals or healthcare centers. However, with the invention of the smart wearable accessories, it has become easier and effective to keep track of basic health parameters like Heartrate and calories. Smart devices such as a smartwatch, wristband, and others keep track of the important vital values. Companies such as *iRythm* are using [21] special stethoscope developed called *EKO stethoscopes* [22] to read the Electrocardiograms values of a patient. These are even delivered at home where the patient or any user can use it themselves. It can amplify the waves by up to 40 times by utilizing various AI-based algorithms.

Figure 9.6 shows the basic use of a health monitoring system. The analytics of the patient is tracked by a device in real time. The data collected can be more than medical data like location or emergency contacts. This data is sent to the doctor for analysis. The analysis step is also sometimes done by Deep Learning Systems. An appropriate decision can be taken based on the report received, and the person can be notified the same.

A Scottish startup Current Health is making use of wearables to record the values of a few vital signs, including pulse rate, oxygen saturation, respiration rate, temperature, and movement, from its own device and others. The company received FDA level II clearance to operate in hospitals. Also, they are already used in the UK and resulted in a decrease in the requirement of home visits by 22% and reduction in hospital readmission rates. It is all thanks to the technology that can monitor the patient in real time and send any alerts to doctors and other healthcare workers. It can reduce the response time by a significant margin and decrease the chances of any accidents.

Wearable devices with added healthcare benefits are great accessories to have and can even help in critical situations for medical needs. A case happened with Bob Burdett when he was to meet his son in a Washington park. He accidentally flipped his

Health Monitoring Devices

Medical Data

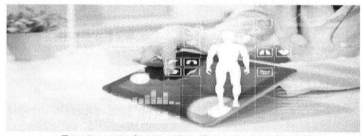

Doctor analyses the data to get insights

FIGURE 9.6 A primary patient health monitoring system

bicycle and hit his head which made him unconscious. Fortunately, he was wearing an apple watch which records any sudden movement changes. His watch detected this sudden movement and regarded it as a fall and sent an SOS signal to his son, who was waiting worriedly for his father. His son called for an ambulance to the location sent by the watch. There is a different kind of systems too that are used for patient monitoring. Some make use of speech data to analyze the mental well-being of a patient. They can alert emergency contacts like relatives or doctors when the patient shows severe stress signals. They can thus help ensure different levels of security for a person.

9.7 CONCLUSION

From all the previous chapters, AI-based systems have brought great changes in our healthcare industry and shown great potential [23]. However, AI is not some flawless technology that can solve all problems without causing any trouble. In fact, it has many of its downsides too. Some of which are mentioned below:

1   It creates a concern for data privacy and security. Since most of the data belongs to some patient, there is a huge risk of data leaks and disclosure of confidential information.

2   It cannot describe the cause and effect of a decision. The amazing powers of AI come at the cost of its interpretability. It is very hard to identify why AI took any specific decision for the given case. Like what feature it focused on to decide that patient has a tumor.

3   There are some ethical issues too. What if the data is used for some harmful purposes and not the one that it was collected for? Also, how can one ensure that there will not be any misuse later?

4   Also, updating hospital facilities is hard as this software requires updates from time to time. This can increase the infrastructure cost and updates cannot be afforded by many healthcare organizations.

5   Also, the meaning of machine learning metrics is difficult to interpret in medical applications as they are hard to grasp for healthcare workers.

But even with all these issues, it can easily be agreed upon that AI has eased our lives and continues to do so with its many applications. It has contributed toward various treatments like cancer, tumor, robotic exoskeleton, and many others. These applications have helped normal users live a better quality of life. Although AI cannot independently solve all the problems with human aid, it can do wonders in many fields. So, it can be safely said that AI is one of the biggest weapons in our arsenal when it comes to conquering the field of healthcare and reach its apex.

REFERENCES

[1]   Narayanan, Krish. "Redrawing the healthcare landscape the facets of AI and robotics." *The Management Accountant Journal* 54.3 (2019): 38–41.

[2]   Rong, Guoguang, et al. "Artificial intelligence in healthcare: review and prediction case studies." *Engineering* (2020).

[3]   McCauley, Nancy, and Mohammad Ala. "The use of expert systems in the healthcare industry." *Information & Management* 22.4 (1992): 227–235.

[4]   Greene, Beth G., David B. Pisoni, and Thomas D. Carrell. "Recognition of speech spectrograms." *The Journal of the Acoustical Society of America* 76.1 (1984): 32–43.

[5]   Chand, M., et al. "Robotics, artificial intelligence and distributed ledgers in surgery: data is key." (2018): 645–648.

[6]   Krizhevsky, Alex, Ilya Sutskever, and Geoffrey E. Hinton. "Imagenet classification with deep convolutional neural networks." *Advances in Neural Information Processing Systems.* 2012.

[7]   Wei, Donglai, et al. "Understanding intra-class knowledge inside CNN." *arXiv preprint arXiv:1507.02379* (2015).

[8]   Nguyen, Dan, et al. "3D radiotherapy dose prediction on head and neck cancer patients with a hierarchically densely connected U-net deep learning architecture." *Physics in Medicine & Biology* 64.6 (2019): 065020.

[9]   Usman, Muhammad, et al. "Motion corrected multishot MRI reconstruction using generative networks with sensitivity encoding." *arXiv preprint arXiv:1902.07430* (2019).

[10] Sohn, Sunghwan, et al. "Clinical documentation variations and NLP system portability: a case study in asthma birth cohorts across institutions." *Journal of the American Medical Informatics Association* 25.3 (2018): 353–359.

[11] Chorowski, Jan, and Navdeep Jaitly. "Towards better decoding and language model integration in sequence-to-sequence models." *arXiv preprint arXiv:1612.02695* (2016).

[12] Hasrul, M. N., M. Hariharan, and Sazali Yaacob. "Human Affective (Emotion) behaviour analysis using speech signals: a review." *2012 IEEE International Conference on Biomedical Engineering (ICoBE)*, 2012.

[13] Budzianowski, Paweł, and Ivan Vulić. "Hello, It's GPT-2 – how can I help you? Towards the use of pretrained language models for task-oriented dialogue systems." *arXiv preprint arXiv:1907.05774* (2019).

[14] Brown, Tom B., et al. "Language models are few-shot learners." *arXiv preprint arXiv:2005.14165* (2020)

[15] Zellers, Rowan, et al. "HellaSwag: can a machine really finish your sentence?" *arXiv preprint arXiv:1905.07830* (2019).

[16] Hinds, William E. "Penicillin product." US Patent No. 2,487,336. 8 Nov. 1949.

[17] Goldsworthy, Peter, and Alexander McFarlane. "Howard Florey, Alexander Fleming and the fairy tale of penicillin." (2002).

[18] Smalley, Eric. "AI-powered drug discovery captures pharma interest." (2017): 604.

[19] Swapnarekha, Hanumanthu. "Role of intelligent computing in COVID-19 prognosis: a state-of-the-art review." *Chaos, Solitons & Fractals* (2020): 109947.

[20] Gupta, Anvita, et al. "Generative recurrent networks for de novo drug design." *Molecular Informatics* 37.1–2 (2018): 1700111.

[21] Bahney, Timothy J., et al. "Physiological monitoring device." US Patent Application No. 14/162,656.

[22] Digger, Kirsty, and Landa Palmer. "Using a Bluetooth Connected Stethoscope: New Technology Teaching Auscultatory Skills in Nursing Students." (2017).

[23] Kelly, Christopher J., et al. "Key challenges for delivering clinical impact with artificial intelligence." *BMC Medicine* 17.1 (2019): 195.

Index

Lightning Source UK Ltd.
Milton Keynes UK
UKHW022357140622
404452UK00004B/35